WIND POWER

Wind Power
Alternative Energy Source

Edited by

Nirmala Rao Khadpekar

2012

Icfai Books
The Icfai University Press

WIND POWER: ALTERNATIVE ENERGY SOURCE

Editor: Nirmala Rao Khadpekar

First Edition: 2012
Printed in India

Published by

This book is published by IUP.
University Campus, Agartala-Simna Road,
P.O. Kamalghat Sadar, Agartala – 799210, Tripura (West)
E-mail: info@iupindia.org
Website: www.books.iupindia.org

ISBN: 978-81-314-2745-3

Contents

Section III

Experiences and Initiatives

Overview

Wind is a cleanly produced renewable energy resource. Wind energy contributes significantly to energy security and good quality environment of living beings. We have a tendency to think a great deal about the quality of life of human beings, but when it comes to the environment, we have to remember that we only share it with the other animals on this globe, and because we are more powerful with our weapons, it does not exclude the quality of life needed by animals and trees and plants. The threat posed by most dangerous weapons of mass destruction where the earth's environment is concerned are probably fossil fuels and their indiscriminate use. Renewable energy resources need to be commercially developed on a scale competitive with conventional sources; which when so described, designate the definition of renewable as unconventional. This should not be acceptable. Renewable should be the first choice and the sheer demand and investment in refining these technologies like sun and wind should bring this about. Also, wind and sun are some of the areas which are the answer to the heretofore electricity denied or deficient for underdeveloped areas of the world where it is difficult for conventional grids to be erected or to reach. Issues that need to be resolved are the lack of wind resource data, remote habitat utility policies, sovereignty, perceived developer risk, limited loads, investment capital, technical expertise, and transmission to markets. The book attempts to look at the economic, technological and environmental aspects of wind harnessing to give a picture of how the wind energy sector has developed and its potential for the future. Information on wind harnessing, wind farming, wind power generation along with the economic, technological and environmental aspects are looked into. For this, the book has been divided into three sections.

Section 1, which is called **Value Proposition**, begins with the Editor's curtain raising article titled **"Wind Harvest: Prospects and Problems"** by *Nirmala Rao Khadpekar* on how, along with being awesome, technology can also be breathtakingly beautiful. At the same time some may view wind turbines as something dangerous that escaped from a science fiction movie set, or even something to have a lance fight with as The Man of La Mancha, the legendary eccentric knight errant Don Quixote challenged a windmill looming on the plains. Coming back to reality, demands for non-carbon forms of energy to mitigate climate change will need some surface landscape altering to save the overall planet from greenhouse gas ruin. Judicious siting and large scale farming of wind is seen as the answer. Further, like in all new technologies the environmental risk assessment has to be meticulously done along with canvassing for government subsidies like feed in tariffs and production tax credits. Any technology costs, so do wind turbines cost but what is the cost that they offset is the enquiry in this article. Investment in wind farms can offset a greater cost to the earth in combating climate change dangers. While noise from turbines can be music to the ears of conservationists, we need to raise the bar for the technology to refine these negatives and treat this elegant form of power with respect and use it wisely.

The second article titled **"Wind Power"** by *Ton van de Wekken* and *Fred Wien* states how wind power has been used over thousands of years for various purposes, but it was the oil crisis which triggered a substantial development and impressive improvements in wind farms are being realized with wind farms now rising out of the sea. It explains the basic principles of turbine wind power generation and goes on to explain how developments including intelligent power electronics in wind power harnessing have continually improved during the last decennia. Considerable scaling up in the wind power industry has taken place with turbines becoming larger, efficiencies and availabilities have been improving and wind farms becoming vast to cope with the increased consumption of electricity worldwide. It tracks how wind power has consistently grown in Europe with Germany, Spain, Denmark and the Netherlands accounting for 84 percent of European capacity. It also discusses future developments, costs and benefits, policy and regulation in the light of environmental issues and lowering of investment and maintenance costs which make this technology interesting for wind farm developers.

The third article titled **"Whose Wind?: Prospects for Cooperative and Community Wind Development on the US Upper Great Plains"** by *Bart Finzel* and *Arne Kildegaard* analyses the potential for the anticipated growth in the wind power industry to be organized cooperatively or at least locally with 'community wind' movements. These represent an alternative form of local ownership to the natural choice of Rural Electricity Cooperatives (RECs). The area under study is the Upper Great Plains of the US which is slated for tremendous wind resource development over the next decade. This region also has a strong reputation in the cooperative movement which came into being over two centuries. The fact that this alternative has a competitive advantage of incorporating and financing themselves in precisely tailored manner to capture greatest tax incentives is examined. The social costs especially of contracting for various wind-development services however are almost certainly higher with community wind than the REC wind. Cooperative wind being slow on the uptake and private wind not having a sharing option, community wind as an alliance between the community and REC wind is suggested for more widely shared community benefits.

The fourth paper titled "**Wind Energy: Issues to Consider**" by *Brian J Frosch and Joe L Outlaw* discusses how though average residential electric prices in Texas have gone up by more than half in over four years; producers are looking to increase capacity ahead of demand. The US alone leads by adding 14,900 MW of wind generated electricity worldwide, surpassing Spain to reach the second overall position. Germany leads the world in wind power capacity. Within the US it is texas that leads. This policy research report is from the Texas Agricultural and Food Policy Centre, The Texas A&M University System as final results of a research project undertaken by their faculty. The article looks at small wind applications for farm and residential use of up to 100 kw generation capacity to Large Wind Applications where landowners in areas with class three or higher winds which measure output in megawatt where each turbine totes up three MW. Additionally Wind Lease Considerations, Wind Lease Glossary, Government Incentives, Availability, Site Selection, Equipment Providers, Utility interconnection, Net Metering and Economic models are touched upon.

The fifth article titled "**Wind Power Today**" is a paper from the US Department of Energy that discusses the plan to provide 20 percent of the country's electricity supply; US wind capacity has to increase from its current 11,600 MW to more than 325,000 MW. Incorporating this in the electricity portfolio could avoid emission of 3,500 million metric tons of carbon equivalent through 2050. This is equivalent to the amount of carbon produced by the entire transportation sector over 3-1/2 years. This would also lead to approximately $332 billion in economic investment and more than 3,725,000 full-time equivalency job years to construction and plant operation, largely focused in rural areas. To address the economic issues faced by rural communities across nations, the work is with rural community leaders, department of agriculture, local and national representatives of Farming interests and the local financial community to explore wind development options, benefits, and barriers.

Achieving the goals of WPA during the next 20 years will create $60 billion in capital investment in rural America, provide $1.2 billion in new income for farmers and rural landowners, and create 80,000 new jobs. Some of the other issues discussed are: Technology acceptance, rural economic development, wind power for native populations, wind for schools, environmental assessment, integration of grid wind R&D, and systems integration. Reducing costs through improved technologies and international collaborations are seen as the answer and projections have been made for the next decade.

The sixth article titled "**Whither Wind?**" by *Charles Komanoff* is a piece of writing which is a philosophical learned presentation on large wind farm siting from a man whose life's work has been about fighting fossil fuels and machines powered by them. The engagement is over the disagreement about whether big windmills belong and whether they belong at all with the backdrop of how the world needs to reduce CO_2 emissions by at least 50 per cent over the next few decades. Based in scenic parts of the US where just to live requires large wattage, the country is taken on as the biggest emitter by far needing to do the job by 75 per cent carbon emission cut to give the poor countries room to develop. The new ethics for wind power siting is discussed in the light of the fact that it should be much easier to comprehend the impact of wind farm development than the less tangible losses from a warming earth. Issues like 'why me, why my ridgeline, my seascape, my viewshed' are examined with a suggested ranking matrix of locales under 'Unsuitable', less than ideal, conditionally favorable, and most favorable to cope with the siting issues. This is where the dilemma of an environment with wind towers or other renewable energy installations could be accepted as 'our generation's way of avowing our love of the next?' to end the argument of an environmental view sullied or no environment at all.

The last article in this section titled "**Wind Energy: A Promising Future**" by *Deepika M G* as compared to solar energy; wind energy is cost-efficient as it has low operating cost, with implicit costs being considered minimal. Wind production cost is roughly five cents per unit while it is 12 cents for solar. Estimates show that both the capital cost and operating cost are likely to asymptotically decline by the next decade due to the likely advancement in technology. The latest advancements in technology and the research being carried out in the field of wind energy include groups working in the US, the Netherlands and Canada to set up wind farms. These installations towering 9 km up in the sky in the tropopause in what is called the jet stream or corridor of high velocity winds hope to garner the 1% of wind energy at high altitudes to meet 'all the energy requirements of the planet'. Australia has worked out how to extract energy using a fleet of whirring gyromills at more than 4 km above the ground. But the whole concept is, however, very challenging as it involves getting the massive windmills which are 7 to 9 km tall, erected and then connect a cable down for use. The installed capacity for India is roughly contributing to only 10% of world capacity. This is expected to grow to a large extent, outdoing other countries as the major wind producing countries in Europe are expected to stagnate in the immediate future due to lower scope for expansion given their small geographic territory with existing technology. In Europe, the capacity is likely to grow only by 17 percent in 2009 and in the US by 23 percent, both being much lower than what is expected in Asia.

Section 2 called **Technology Considerations** begins with an article titled **"A Review of NDT Techniques for Wind Turbines"**by *M A Drewry* and *G A Georgiou* reviews the current state of Non-Destructive Testing of wind turbines at manufacture and in service in Europe and internationally where wind turbines are required to generate high quality network frequency electricity. It also attempts to establish the most promising NDT methods (UK based) for detecting flaws of most concern because mass production of wind turbines is fairly new and no manufacturing standards have as yet been set. The historical 'Dutch' windmills which preceded electricity for grinding grain were used in Persia in 200 BC and have been in use in turbine form since 1941 when the first MW size was connected to a local electricity generation system in Vermont, USA. The paper reviews the NDT of wind turbines based on published evidence, at the time of manufacture and at the point of periodic inspection, including Acoustic Emission which does not strictly come under NDT. It concludes with the need to develop European standards for use by developers, investors and insurance companies on risk, economic viability, performance, reliability, and owners and managers of wind farms.

The next article is titled "**Offshore Wind Power on the Horizon: A New Energy Frontier for Oceans, People and Wildlife**" by *Jeremy Firestone, Sandy Butterfield, Christina Jarvis* and *Jack Clarke* which discusses how offshore wind holds much promise as a means to meet electrical demand in coastal states, particularly if potential sociocultural and environmental impacts are thoughtfully considered and addressed. Offshore wind is viewed from the perspective of large wind resources that exist above the Atlantic Ocean, close to the east coast cities. With several manufacturers developing large wind electric turbines designed to be placed offshore in waters up the 30 m in depth, offshore wind technology and potential is discussed for offshore wind development off of Cape Cod and Long Island. Impacts and benefits of offshore wind development on humans and wildlife are discussed along with factors underlying public opinion on offshore wind. Floating platform systems at some deep sites for economy and feasibility are a reality along environmental changes like habitat disruption due to underwater noise and vibration with even possible habitat enhancements with the provision of artificial reefs.

The next paper is titled "**Wind Farm Development and Operation: A Case Study**" by *Ton van de Wekken* which tracks how implementation of wind energy has changed dramatically over the last 20-25 years from stand-alone to major wind farm development with turbine varying from a few to as more than 25. Wind farms with established capacity of 50 MW are not uncommon and wind turbines in the range of 4.5 to 6 MW are already available on prototype. Big is becoming beautiful. Noticeable tendency to go offshore for siting has been noticed though the investment costs are double as compared to onshore, not consider the high expenses incurred for sea foundation, sea cable and special sea vessels for transport and erection of equipment. Despite the operational costs also being at least double as compared to onshore, the nuisance to habituated environment is low and excellent wind resources lead to a high degree of utilization. For most locations, wind energy needs the perquisite of incentives for using renewable energy and not emitting carbon emissions to be profitable because wind farms cannot be cost-effective on the investment and ROI parameter alone.

This is followed by a paper titled "**Impact of TradeWind Offshore Wind Power Capacity Scenarios on Power Flows in the European HV Network**" *J O G Tande, M Korpås, L Warland, K Uhlen* and *F Van Hulle*. This paper examines the simulations which have been carried out for studying the effect of offshore wind on power flows in the European high voltage network with a detailed model including onshore wind capacity for 2005 and 2020. Wind power data from the Re-analysis global weather model, combined with regional wind power curves and wind speed adjustment factors is used in the simulation for the different grid zones. The analysis shows that bottlenecks between countries exist but predict reduced costs for 2020 scenarios with a modest cost attribution to wind variations. Actual bottleneck costs experienced by system operators are expected mainly due to market constraints for trade between countries, and due to country limitations in transfer capacity. 'Trade wind' is the name of the project initiated by the EWEA to investigate the necessary interconnection upgrade and electricity market rules to facilitate the integration of the future expected amounts of wind power.

Section 3 titled **Experiences and Initiatives**, has four papers, the first one being titled "**Support for Wind Resource Assessment and Wind Farm Development in China**" by *W L Wallace, Z Y Wang* and *S J Liu*. This paper reviews the support for resource assessment and trends for wind development in China and updates about the National Wind Concession Program. National planning targets for China and policy support for commercial development from the point of Renewable Energy Law are examined. Since 2000 the UNDP and Global Environment Facility have cooperated with the country's National Development and Reform Commission to pursue a wind resource assessment and analysis program to proliferate the technology of wind power generation. The current state of development of Wind in China is looked at along with future prospects, support for risk assessment and site characterization, selection, equipment procurement and installation, field configuration, data analysis and site development. Scrutiny of quality and reliability of data and its analysis to support an increasingly receptive investment environment is sought to meet the aggressive targets for wind power development in China are outlined.

The next article is titled "**Institutional Factors and Knowledge Co-Evolution for Developing the Wind Energy Industry in India**" by *Rekha Rao*. Here the authors explore the factors that promoted the development of nascent wind energy sector in India leading it to become one of the important wind energy producers in the world. They investigate this aspect with respect to India's unique policy of joining hands with other countries with advanced wind energy sector to develop its own wind energy industry. In this study we concentrate on the relationship between India and Denmark. The institutional factors responsible for development and sustenance of the wind industry in India are highlighted along with information on technology policy initiatives. Their analysis leads them to believe that knowledge co-evolution within and between institutions related to wind industry was responsible for helping India establish itself as the fifth largest wind energy producer in the world. The authors feel that their attempt to bridge the gap in the knowledge of how the wind energy industry developed in India will provide insights to the institutions and actors in the wind energy sector.

This is followed by the article titled "**Wind Energy in India: The Future and the Challenge**" by *A Lakshminarasimha* and *S Rath* which discusses how with a fifth in the world ranking in energy consumption, India needs to accelerate the development of the wind energy sector to meet its growing needs. It looks at the potential fiscal and financial incentives in the current scenario in India with challenges faced by the wind sector and the next steps to be taken to realize success. Barriers to accelerate growth are examined with future projection for wind development in India with variable to illustrate the importance of their influence of to the growth of wind industry in India. Further, by analysis of these, three different scenarios are drawn up based on the macro and internal factors that will be mainly responsible for the growth of wind power in India in the years to come – Optimistic, Neutral and Pessimistic which are further analyzed and presented. The authors predict that their analysis would help upcoming wind energy promotions in a situation where there is non-alignment between the demand need time and electricity supply available time when it is based on wind energy.

The next paper is titled "**Research and Assessment of Wind Turbine's Noise in Vydmantai**" which is located in Lithuania by *Bronius Jaskelevičius and Natalija Užpelkienė*. This place has a seaside most suitable for wind energy development and a health resort. An environmental debate has emanated due to noise from the turbine machinery disturbing the activities. Fighting climate change with renewable energy has to trade off effectively with the fallouts that may disrupt or create a different set of problems for human beings that may not measure against global warming. Distrust has to be addressed and caution utilized to offset negative effects on people like noise and view disruption, posing a danger to flying birds and harming local soil biocenoses. Terrain also has a great impact on noise proliferation of wind power generation as shown in this study for which science and advanced technologies are suggested for mitigation and selection of suitable wind turbine installations. The point is that technologies for improving the engineering are possible and should be looked into and indiscriminate intrusion based on one parameter at the cost of another should be avoided.

The last article of the book is a case study called "**Wind Resources Inc.**" by *Rocky Higgins and Robert Keeley* from UW Business School. It is about the proposition to build and operate a 30 megawatt wind energy facility on the site of a company's easement to develop the Canyon Wind site located in the San Gorgonio Pass near Palm Springs, California. There was a complex 20 year agreement with private investors, creditors, and the Southern California Edison Company which posed a dilemma as private investors had encountered difficulty in servicing the debt originally incurred to purchase the wind turbines. Market wisdom indicated a continued use and upgradation of the existing wind energy equipment to much larger turbines. The original agreement about to expire, the CEO of the company in question was trying to decide how best to capitalize on the company's easement to develop this easement when the shareholders were anxious to cash out their investments as soon as possible. The case also deals with how the value of the project was estimated for fixing a minimum price for any possible sale or auction.

As an annexure, we have included a paper titled "**Guidelines for Conducting Bird and Bat Studies at Commercial Wind Energy Projects**", which looks at the fact that while wind energy has significant environmental benefits when compared to energy produced from fossil fuel, the potential environmental impacts of both site specific and cumulative impacts that commercial wind energy must be considered while evaluating proposed projects. This paper outlines the model guidelines of the US Department of Energy Conservation to commercial wind energy developers on how to characterize bird and bat resources at onshore sites and how to estimate and document impacts resulting from construction and operation of projects. The purpose of the document is to outline the protocols for conducting bird and bat studies to provide information to the DEC to assess ongoing or expected environmental impact to enable them to recommend construction and operation requirements to avoid or minimize adverse environmental impact. Along with the study objectives and rationale, standard, and expanded pre- and post-construction studies are recommended along with bird and bat mortality analysis and reporting with a provided valuable list of sources of information which includes the renowned Audubon Society.

Conclusion

The book attempts to put together valuable material on the subject to emphasise that such interventions on renewable energy need to be addressed intelligently. Climate change occurring due to indiscriminate burning of fossil fuels is not just an environmental threat; it is an economic aspect threatening the planet. Technology can be beautiful, if only we can design it so, to let it gracefully and sensibly come into our lives by planning for it, investing in it and refining it as required for enriching our lives. Right now, we have to save our lives from ruin. Let the renewable energy forms be the symbols of elegance and beauty and not be halted by varying price of fossil fuels which have hidden environmental costs that we preferred not to see for many years. We need to move rapidly towards supporting a widely distributed proposition like harnessing the wind.

Section I

Value Proposition

1

Wind Harvest: Prospects and Problems

Nirmala Rao Khadpekar

One way or the other, renewable energy, alternative energy or non-conventional energy has to be on top of the mind as the raves of conventional fossil fuels need to be halted. Wind energy is a pollution-free, substantially sustainable type of energy which uses no fuel, produces no greenhouse gases; neither nuclear type of waste nor toxicity. Wind turbines today are mammoth and numerous, making wind farming and wind harvesting serious options for the electricity industry and are even supplied to grids. Smaller turbines can provide electricity at isolated locations and individual installations. Lands siting and environmental risk assessments require to be sensitively and judiciously executed, and technology needs to be welcomed as awesome grandeur and beauty for the betterment of the environment for future generations.

Introduction

In a perfect world, one full of excellent choices mankind could simply preserve the pristine beauty of the landscape and make sure that no technological interventions sully the environment. But, when the choice before us is burning fossil fuels and eventually not having any environment left, sustainable energy choices offering renewable power like wind energy will need to accept technological intrusions into the environment. While technology can be awesome, it can also be breathtakingly beautiful. Wind turbines can be seen as a science fiction movie set, or even something to have a lance fight with as Don Quixote did with a windmill in the legendary play *The Man of La Mancha.*[1] Row upon row of tall slim wind turbines, dot many landscapes... some find them awesome and others resent them as interfering with the scenic beauty of the certain environmental locales like sea shores and mountain ridges which are rich wind harvesting sites. The newer turbines are mammoth, larger than the Statue of Liberty. At the end of 2007, global wind power generation is said to have increased more than five fold in seven years.

What is wind after all, it is moving air, it is a strong breeze, it is produced by the movement of the sun,[2] it is worshipped, it is feared, it can cause damage, literature in all languages have a wonderful and beautiful body of rich phraseology describing the power and the glory of 'Pavan'. What wind energy does is to convert energy resulting from motion that is there in wind into more useful forms of energy such as motorized energy or electricity. Wind energy is a pollution-free, substantially sustainable type of energy. It uses no fuel, produces no greenhouse gasses; neither nuclear type of waste or toxicity. It is a reverse process; we generally use electricity to produce wind with a fan, while here it is the wind that turns the fan in which the turbine uses that wind energy to generate electricity for a host of other purposes.

Historical Background

Human beings have used wind energy for thousands of years. Ancient Persians before the birth of Christ used wind energy to pump water. Early windmills used wind energy directly as mechanical energy to grind grain or pump water. Today wind turbines generate electricity at major windfarms which are supplied to electric grids. Smaller turbines are used to provide electricity at isolated locations. Man has been harnessing the wind for many thousands of years which have a BC qualifier. Wind-powered sailing ships explored the world for hundreds of years before the

---

1 *Man of La Mancha* is a musical adapted from Wasserman's non-musical 1959 teleplay I, Don Quixote, which was in turn inspired by Miguel de Cervantes's seventeenth century masterpiece Don Quixote. It tells the story of the "mad" knight, Don Quixote, as a play within a play, performed by Cervantes and his fellow prisoners as he awaits a hearing with the Spanish Inquisition. *Source: Wikipedia Entry.*

2 *encarta.msn.com/encyclopedia_761562112/sun.html*

invention of first steam engines and then fuel engines and electric ones. Sailing ships are being revived for transporting cargo.[3]

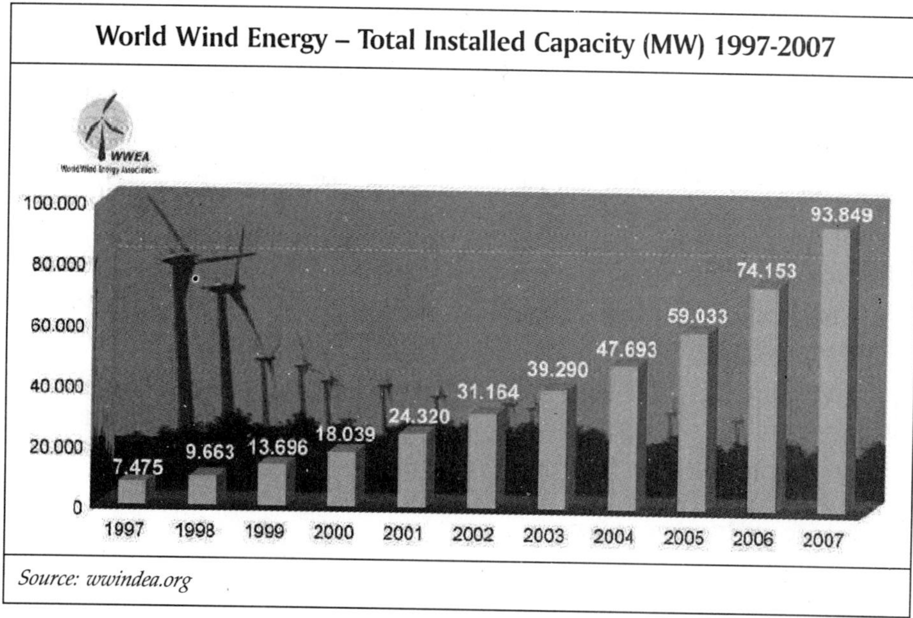

World Wind Energy – Total Installed Capacity (MW) 1997-2007

Source: wwindea.org

Human beings have used wind effectively over millennia. The original heyday in America was between 1870 and 1930, according to a paper on *How Wind Energy Works*, in 'Clean Energy' from the Union of Concerned Scientists (Citizens and Scientists for Environmental Solutions). From pumping water and grinding grain, small electric wind turbines were soon thereafter put in use by the 1930s and by 1940 the farmers in rural areas had the prototypes of larger contraptions. The New Deal began with supply of grid connected electricity to the countryside; windmills began to be phased out. In Europe, windmills have been part of the scenery for centuries. In fact, some Scandinavian countries are identified with the motif of windmills (and wooden shoes in folklore) and they are making history with new technology.[4] Even the local Greenpeace office in Copenhagen, owns shares in the wind power projects. For details on Denmark's achievements in renewable energy, notable wind power, see *Denmark's Windmills Flourish as Cape Cod* power project stalls, by Charles M. Sennott, Staff Member, *The Boston Globe.*

3 *www.treehugger.com/files/2008/03/beluga-skysails-cargo-ship-kites.php-113k*

4 Vestas.

Wind Power; Top 10 Countries, 2005 and 2006

	Country	Existing in 2005 MW	Added in 2006 MW	Existing in 2006 MW
1	Germany	18,420	2,230	20,620
2	Spain	10,030	1,590	11,620
3	United States	9,150	2,450	11,600
4	India	4,430	1,840	6,270
5	Denmark	3,120	10	3,140
6	China	1,260	1,350	2,600
7	Italy	1,720	420	2,120
8	United Kingdom	1,330	630	1,960
9	Portugal	1,020	690	1,720
10	France	760	810	1,570

Source REN21, Renewables 2007, Global Status Report.

Technology Interventions

There is an estimated 72 TW (Terra Watt) of wind energy across the world which is considered commercially viable as it has to be understood that not all the wind that blows can be captured. Wind energy is estimated to generate 10-20% of the world's electricity by 2030. Wind produced energy is looked upon favourably by conservationists as an option to something you have to dig from the ground. It is abundant, does not get diminished, is widely distributed, uncontaminated, and low on greenhouse gases. Yet, construction of wind farms is not unanimously welcome due to their sheer vastness on environmental impacts. The irregularity of the wind blowing rarely causes problems in the case of utilizing wind power to supply a small percentage of full quantity of the requirement.

At large installation levels, wind energy is produced by installation of large three bladed wind turbines and operates like fans in reverse motion. Starting from 20 m rotor blade diameter in the 1980s, a rotor blade (with hub) can be 90 m today, allowing for capture of four times more energy.[5] Wind speed studies track, how wind as a resource track, how fast it blows and how often, and when, which play a critical role in the cost of wind generation. Such studies can take up to two years of readings.

[5] National Wind Technology Centre as told to America.gov; *http://www.windsolutions-europe.com/PDFs/fl2500-tech.pdf*

The power garnered from a turbine arrives as a 'cube of wind speed'; which means that if the wind speed doubles, there is an eight fold increase in power generation. Anemometers are used on meteorological wind towers to measure wind resources. A less reliable way is to use survey maps and readings collected by local met offices. The table here shows the seven classes of wind that can be available from a wind resource study and 'the turbine matched to the speed and frequency of the resource to maximize power production'.[6]

Classes of Wind Power Density at Heights of 10 m and 50 m[(a)]

Wind Power Class	10 m (33 ft)		50 m (164 ft)	
	Wind Power Density (W/m^2)	Speed[(b)] m/s (mph)	Wind Power Density (W/m^2)	Speed[(b)] m/s (mph)
1	0	0	0	
	100	4.4 (9.8)	200	5.6 (12.5)
2	150	5.1 (11.5)	300	6.4 (14.3)
3	200	5.6 (12.5)	400	7.0 (15.7)
4	250	6.0 (13.4)	500	7.5 (16.8)
5	300	6.4 (14.3)	600	8.0 (17.9)
6	400	7.0 (15.7)	800	8.8 (19.7)
7	1,000	9.4 (21.1)	2,000	11.9 (26.6)

Source: Energy Information Administration.

As for turbine design, the most common style is said to be the 'horizontal axis design' large or small. Two or three blades spin on the top of the tower it is placed on. Smaller wind turbines like at old farms have tail fans that adjust to the orientation of the blowing wind. Large turbines have hydraulic controls that adjust the blades as required. Prime spots which are high wind are farmed for wind with hundreds of turbines which can provide power to several homes.

Siting and Environmental Issues

Judicious siting of wind turbines becomes a critical issue in wind harnessing, as land will generally be agricultural land, urban agglomerations or ecologically sensitive land, all of which have some siting fallouts. The area alienated by the towers of a

6 How Wind Energy Works, Clean Energy.

wind farm are said to be less than 1 per cent of the area spanned by the wind farm.[7] Wind farms are often located on cleared farming land with the farming activities continuing on the land. In fact, wind generation amounts to a second cash crop to farmers and provides a revenue model to stay on the land in difficult times. A large fraction of wind farms of the future will be located offshore to benefit from the higher wind velocities and to avoid land use conflicts. These sitings will of course be cost-intensive and need special underwater building technology. It would be the same as those for oil rigs and sailing platforms in deep port areas and where there are strong and choppy sea currents caused by high wind. The most scenic sites often have the best wind, and approximately a fourth of the proposed wind installations are delayed due to landscape concerns and wildlife worries.

Noise from wind turbines can be considered music to some discerning ears; those that understand the need for technology to battle climate change issues. Resentment of noise is understandable, especially when it is constant and not tolerable. The technology bar needs to be raised to refine the negatives and promote this elegant form of power and use it wisely. Siting of wind plants/farms and or even just a turbine, can arouse concerns within affected communities. While promoting energy security for the long run, which is already now, some environmental concerns also have to be taken into account. While wind energy offers ecosystems benefits, there is negative fallout on indigenous wildlife habitats and bird species. Just as there is environmental impact assessment done for industrial projects, so also should it be done for renewable energy installations, with its own parameters. While the acceptance is growing, proliferation of these large installations is a cause of concern to human beings as well. Public perceptions about the renewable energy industry, regulatory requirements in relation to environmental issues have to be taken up along with the costs. Such environmental impact assessments have to be seen against the coal mining disturbances and its impact on the land and life. Chapter 5 in the book 20% wind by 2030 estimates that there are considerable water savings and atmospheric cleanliness involved as compared to mining. Wind harnessing projects utilize the same land unchangingly, while coal has to shift to new sites after exploiting and degrading the currently used site.

---

7 *www.garnautreview.org.au*

Bird and bat impact studies are numerous and give practical guidelines on how to go about such large wind capture installations. And, though wind energy may be able to thrive along with farming, cattle rearing and forestry activities, it is far more complex to get the compatibility with housing settlements, airport accesses, and low flying aircraft paths, both army and civilian. Direct impact of building turbines, like service roads, substations can cause displacement of up to 0.4 hectare per turbine. Indirect impact can be from tree pruning forest being degraded from the point of view of viable habitat loss foe birds.[8]

The answer to addressing these issues seems to be going offshore, which can be done when there is huge investment potential available. Offshore offers the advantage of strong winds blowing over the surfaces of oceans or large lakes. While there are no land use issues offshore, the laying of deep undersea cables and preventive/repair maintenance teams to keep the turbines going in sometimes extreme weather conditions is not escapable.

Dealing with the Economic Threat

Climate change occurring due to indiscriminate burning of fossil fuels is not just an environmental threat; it is an economic threat[9] and needs to be addressed intelligently. Green energy and energy conservation propagators are beginning to see wind turbines as symbols of an elegant form of power to make that transition as we need to move as quickly as possible to a widely distributed activity like harnessing the wind. Wind power is a capital intensive technology, requiring about 75 per cent of the costs right away.

Wind turbine technology costs but they also offset environmental costs of great climate change dangers. There are three key drivers which are forcing markets towards renewable energy options. The first is energy security for the country by refusing dependence on foreign oil despots. The second driver is of course the concern about climate change to decrease greenhouse gas emission from burning fossil fuels by use of carbon-free green energy. The third driver is the steady cost decrease in renewables due to refinements in technologies, especially in Wind Energy. In the US, the key to maintaining the industry's momentum is to extend the tax credits and make them

8 Wind Power citing and Environmental Effects, 20% Wind Energy by 2030.

9 Largely sourced from 'How Wind Energy Works, Clean Energy'.

more stable for wind production according to Randal Swisher, Executive Director of the American Wind Energy Association. America has been the largest wind turbine market since 2005, and has a large stake in promoting this technology.[10]

There have to be workable systems put in place to encourage the investment in renewable energy which are separate for carbon reduction rewards. The terminology used for subsidies vary in different countries – feed-in tariffs, green certificates, tax rebates or tender in addition to carbon credits. Various mechanisms after the Kyoto protocol are straining to be in place and need to help investors in renewables to get on with their plans. Clean Development mechanism and Joint Implementation issues have "project-based schemes" that "play a role in the development of wind energy projects in third countries, but they do not constitute the decisive element for a company that is considering investing there.'[11] But it is the impact of the local body structure and the long run stability of the carbon markets which is quite critical. Their effect on wind energy investments would increase if problems like the shortening of the procedures and the review of the certain clauses were organized, say the authors of the paper. India, in the paper, "Wind Energy in India: The Future and the Challenge", by A Lakshminarasimha and S Rath in *The Icfai Journal of Infrastructure*, Vol. V, No. 4, 2007, says that the benefits for wind entrepreneurs worked like this:

- Concessional import duty on specified wind turbine parts;
- An accelerated depreciation of 80% in the first year;
- Sales tax, excise duty reliefs;
- Loans through Indian Renewable Energy Development Agency (IREDA);
- Income Tax holiday; and
- Nine states have introduced these policies.

Bird's Eye View

In many nations wind has attained comparatively high levels of reach in 2007, amounting to about 19% of power produced in Denmark, with as much as 9% in

---

[10] *http://mumbai.usconsulate.gov*

[11] Can carbon markets support wind energy investments? María Isabel Blancoa, Glória Rodrigues Department of Economic Analysis, Faculty of Economics, University of Alcalá, Plaza de la Victoria 3, 28002 Alcalá de Henares, Madrid, Spain. European Wind Energy Association, EWEA, Rue D'Arlon 63-65, 1040 Brussels, Belgium.

Wind Power: Top 10 Countries 2005 and 2006				
	Country	Existing in 2005 MW	Added in 2006 MW	Existing in 2006 MW
1	Germany	18,420	2,230	20,620
2	Spain	10,030	1,590	11,620
3	United States	9,150	2,450	11,600
4	India	4,430	1,840	6,270
5	Denmark	3,120	10	3,140
6	China	1,260	1,350	2,600
7	Italy	1,720	420	2,120
8	United Kingdom	1,330	630	1,960
9	Portugal	1,020	690	1,720
10	France	760	810	1,570

Source REN21, Renewables 2007, Global Status Report.

Spain and Portugal, and only 6% each in Germany and the Republic of Ireland. Estimates of the social costs of wind power for countries including Germany, Denmark, Spain, and Ireland by analyzing their electric systems in a standardized setting with and without wind input have been conducted.[12] Germany has an installed capacity of 20 GW plus a yearly production in the year 2007 of approximately 40 TWh, according to Prof Hannes Weigt, Chair of Energy Economics and Public Sector Management of the Dresden University of Technology, Faculty of Business and Economics in the paper "Germany's Wind Energy: The Potential for Fossil Capacity Replacement and Cost Saving."[13] India is one of the few developing countries which have taken a lead in the development of renewable energy from as early as the 1970s. India recognized the importance of natural sources of energy and established a Commission for Additional Sources of Energy in 1981 after which the sector witnessed a slow but steady growth for the next three decades of various policy initiatives and economic support interventions.

They are also being integrated into the mammoth building design proportionately. Three vast wind turbine blades admeasuring 29 metres in diameter have been successfully installed on the Bahrain World Trade Center, a twin skyscraper complex

[12] Boccard (2008).

[13] *papers.ssrn.com/sol3/papers.cfm?abstract_id=1262782*

in 2007 a first for a commercial development; it has 'integrated large-scale wind turbines within its design to harness the power of the wind.'[14] They are supported by bridges spanning between the towers. The land Gulf breeze is channeled into the passageway of the turbines through its positioning and the unique aerodynamic design of the towers, helping to create power generation efficiency. These turbines are expected to deliver the equivalent of a 300 homes' annual power requirement, but it will be only approximately 15 per cent of the buildings needs!

Conclusion

As a cleanly produced renewable energy resource, wind energy contributes to energy security and environmental quality. Wind resources could be commercially developed to provide electricity and revenue to underdeveloped areas. Issues that need to be resolved are the lack of wind resource data, remote habitat utility policies, sovereignty, perceived developer risk, limited loads, investment capital, technical expertise, and transmission to markets. Wind energy is considered a clean energy before and after it is produced. It has immense possibilities for developing countries. One of the most inspiring sights to see was in an article by Thomas Friedman,[15] an electricity pole in an urban area with wind turbines at the top with solar panels at mid level. How much more technological amazement can we hope to see in this polluted world? Friedman says, "This will require fundamental reshaping by government of the prices and regulations and research and development budgets that shape the energy market." While renewable energy is not a silver bullet, it is a maturing area for technology that needs to be taken further; it is one of those propositions that improve with use, as it reduces oil imports and consumption, reduces pollution and carbons emissions and, best of all one single major positive, increases jobs. Investment begets returns manifold. Non-harm to the atmosphere consumption spawns new economic activity and becomes a market driver putting clean money back into the pockets that are investing. What better deal can there be in this complex world!

(Nirmala Rao Khadpekar, Senior Faculty Member, the Icfai Research Centre, Ahmedabad. The author can be reached at nirmal.khadpekar@gmail.com, nirmalak@iupindia.org).

[14] *www.treehugger.com/files/2007/03/bahrain_install.php*

[15] "Open to the Sun and Wind", August 14, 2008, *Indian Express.*

Wind use in Literature

There are many phrases regarding the word 'wind' which come along with a history. Much of it is to do with sailing and even imaginary battles. *Three sheets to the wind* and *Whistling in the wind* are sailing phrases, while *Knowing which way the wind blows* means judging the tide of opinion and *an Ill wind blows nobody any good* means exactly that. *Getting wind* of something is to hear a rumour, while *put up the wind* means alarm or frighten and sail close to the wind, verge on indecency, dishonesty or disaster; *take the wind of someone's sails* is to frustrate someone by preempting. And Milton has freely phrased up with the word 'wind' in Paradise Lost.[16] The legendary Don Quixote was reputed for *titling the windmills* – an iconic scene from the book where he thought the windmills were thirty to forty hulking giants rising from the plain to attack him. The phrase derived a good 150 years later to mean unequally prepared for an imagined battle. *The answer is blowing in the wind* is the famous 60s Bob Dylan moving protest song, which incidentally has no answers, only more questions…

2

Wind Power

Ton van de Wekken and Fred Wien

For thousands of years, wind power has been used for various purposes. Mainly since the oil crisis, a substantial development has taken place and impressive wind farm projects have been realized. Over the last few decades the development and commercial production of wind turbines has been active and well-stimulated. Turbines are becoming more efficient, power rates are increasing and intelligent power electronics are being introduced. In the meantime, impressive wind farms are rising out of the sea. Not only environmental issues make wind turbines more often seen on the horizon. The continuously lowering of the investment and maintenance costs of wind turbines make this technology interesting for investors and wind farm developers.

Introduction

What is Wind Power?

General

Wind turbines make a major contribution to the production of renewable energy. When the oil crisis occurred in the 1970s in Europe, the development and commercial production of wind turbines for generating electricity was strongly stimulated. Developments in harnessing wind power continually improved and during the last decennia a considerable scaling up has taken place in the wind power industry. Turbines have become larger, efficiencies and availabilities have improved and wind farms have become bigger.

The consumption of electricity keeps growing on a worldwide basis. Most European governments have set targets to reduce the emission of carbon dioxide in order to stop the Earth from warming up further. The widely accepted opinion is that these targets can only be met on the one hand by energy-saving incentives and on the other hand by the large scale application of renewable energy.

The use of wind turbines is a serious alternative for achieving these aims. Several European countries have impressive plans for the coming years for installing large amounts of wind power generation. Some governments support these actions by providing tax or investment incentives. The northwest of Europe with its coastal windy waters and finemeshed but strong electric grid give promising opportunities for investing companies and wind farm developers.

Basic Principle

Wind turbines extract the energy from the wind by transferring the thrusting force of the air passing through the turbine rotor into the rotor blades. The rotor blades are aerofoils that act similarly to an aircraft wing; this is the so-called principle of lift. This can be seen in the cross-section of a rotor wing in Figure 1.

Figure 1: Cross-Section of a Rotor Wing Showing Speeds and Directions

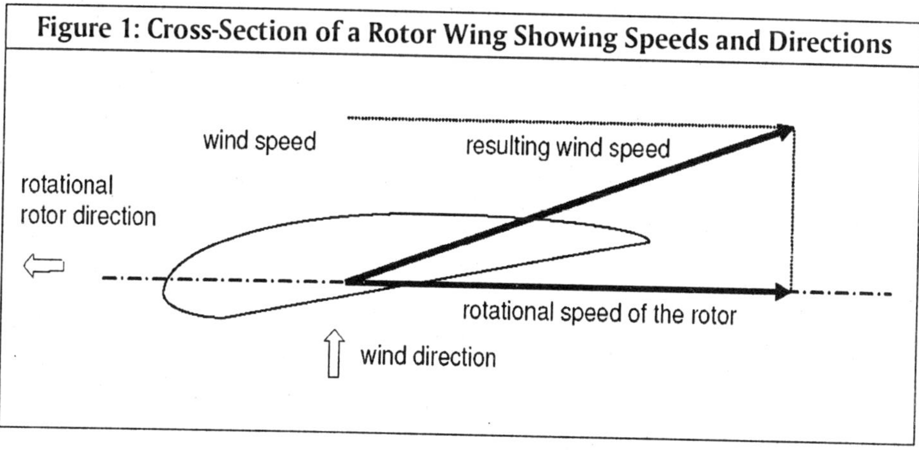

As an effect of the resulting air flow, the windward side of the aerodynamic profile is over-pressured while the leeward side is under-pressured. This differential pressure creates a thrust force. This lifting force is perpendicular to the direction of the resulting force (resulting wind speed) reacted by the flowing wind towards the turbine wing and the local rotational speed of the wing. As a result, the lifting force is converted into a mechanical torque. The torque makes the shaft, as part of the turbine rotor, turn. The power in the shaft can be used in different ways. For hundreds of years it was used for the grinding of wheat or pumping water but the large machines of today, with integrated generators, convert the shaft power into electricity.

Power and Efficiency Rates

Mass in motion carries a certain amount of energy. This kinetic energy varies in proportion to the product of the mass and the square of the velocity. In units of time, this energy is similar to the power.

Kinetic energy per second is:

$$P = 1/2.m.V^2 \quad (1)$$

P power (Nm/s or Watt)

m air mass per second (kg/s)

V wind velocity (m/s)

This physical law is also applicable to air in motion. The mass of air flowing through the rotor has to be imagined as a cylindrical disc. The volume of the disc is equal to the surface area of the rotor and the length of the disc is equal to the wind speed.

Air mass through the turbine rotor per second is:

$$m = \rho.A.V \quad (2)$$

ρ density of air (kg/m^3)

A rotor surface area (m^2)

V wind velocity (m/s)

This leads to an important characteristic; the amount of energy is similar to the speed of wind raised to the third power.

Both formulas combined make the theoretically power:

$$P = Cp.1/2.\rho.A.V^3 \quad (3)$$

Cp mechanical power coefficient (at slow shaft)

As an example:

At a wind speed of 6 m/s the energy content is 132 W/m^2 (Watt per square meter). When the wind is blowing at a speed of 12 m/s, the energy content is 1053 W/m^2. Summarized: twice as much wind speed gives eight times as much power.

Not all the energy present in the wind can be converted into usable energy at the rotor shaft. Using physical calculations it can be proven that the theoretical maximum efficiency of wind power is limited at about 59.

The net electrical power output of a turbine can be determined when mechanical and electrical performance rates are also taken into account.

Net electrical power output:

$$P_{elec} = Ce\ .1/2.\rho.A.V^3 \quad (4)$$

Ce electrical efficiency rate (%)

Today, large modern turbines are able to achieve a total net efficiency *Ce* of 42 to 46% with respect to the energy of the undisturbed wind in a circular tube with a cross-sectional area equal to the gross rotor area.

Basic Comparison with Conventional Electricity Production and Benefits of Wind Power

There are several reasons for the success of wind power over recent years. When compared to conventional electricity production, wind turbines produce clean energy. In operation there is no carbon dioxide emission or other air, water or soil pollution.

Other advantages of "Wind power fuel" are that it is free, it is inexhaustible and it is abundant. Wind turbines have a modular concept (reference to Section 8.4.2 "Standards" of this application note), are quick to install and are also independent of the importation of fuels and fluctuations in fuel prices caused by oil politics. Looking at the reliability of a modern wind turbine, the operating availability (the proportion of time in which a windmill is available to operate) is about 98%. No other electricity generating technology has a higher availability rate.

A disadvantage of wind power is the unpredictability of wind. Storm fronts in particular can cause a sudden increase in the wind power. Furthermore, periods of low wind give little wind power.

The expansion of a grid with power produced by wind turbines is not as obvious as it seems. A certain percentage of the total generated power still needs to be provided by (central) "stable" conventional plants. This percentage depends on the composition and stability of the grid. If the instability of a grid is foreseen, it can be prevented by using intelligent control systems that interface between different kind of production units, consumers and the intermediate grid. In many EU countries, grid companies or (independent) associations are carrying out research into this.

Application of Wind Power

Description of Typical Situations in which Wind Power can be/is Suitable

The amount of electrical power produced by a turbine depends on the size and type of the turbine and where the turbine is located. A characteristic that represents a typical power output in relation to the wind speed is given in Figure 2. At low wind speeds, no electrical power is generated. From 2 Beaufort (approx. 3 m/s) and above the turbine is rotating, and at about a wind force 6 Beaufort (12-13 m/s) the turbine is supplying its maximum power.

At wind speeds over 25 m/s former generations wind turbines were designed to shut down in a controlled way to avoid overloading or damaging the turbine's installation or construction. Modern turbines however are equipped with a pitch control that changes the angle of the rotor wing in extreme weather. The result is that the power supply can be guaranteed even in bad weather conditions. When very heavy storms occur it still is necessary to stop the turbine.

Figure 2: Typical Turbine Characteristic; Power Output in Relation to Wind Speed

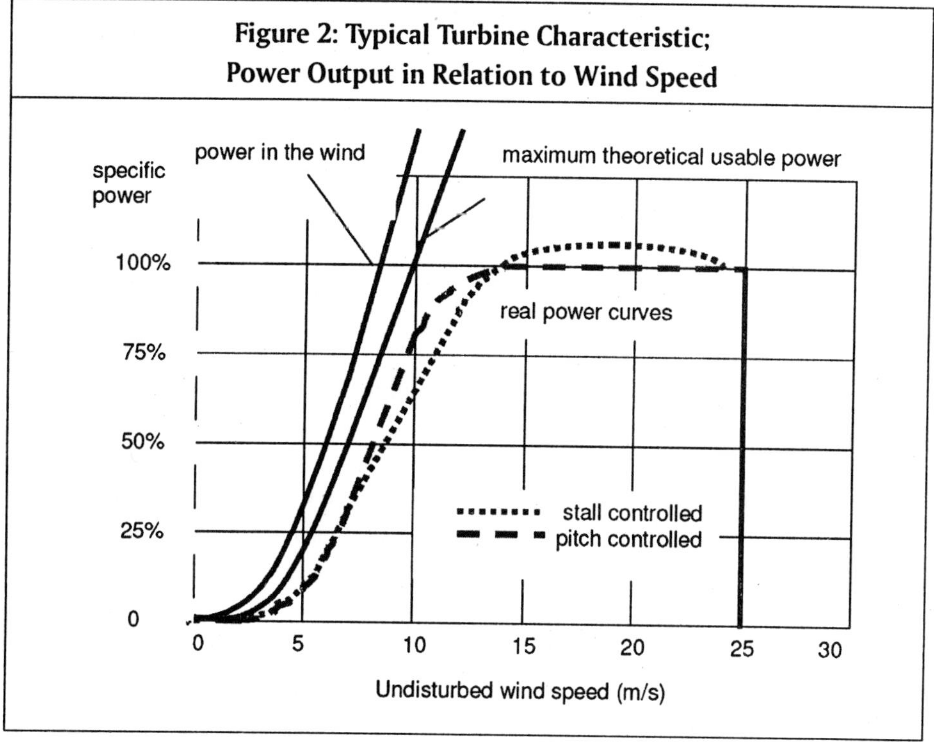

An average turbine in an ideal location can deliver an electrical power output on an annual basis of about 850 kWh per square meter of rotor surface area. Another simple rule for estimating the annual energy yield of a wind turbine is that on an average wind site the output is about 2000 full load hours and at high wind sites approximately 3000 hours.

For example, on average a 1.5 MW wind turbine produces 3.106 MWh, consisting of 2000 hours at 1.5 MW.

Costs of Wind Energy

Without taking into account tax benefits or other incentives on production of wind energy, the costs of wind energy are summarized in Table 1.

Included in other operational expenses are: daily management, insurance, land lease, compensation for visual or noise nuisance and taxes.

Table 1: Summary of Wind Energy Cost Breakdown

Wind Energy Cost Breakdown	2000 Full Load Hours [EUR/MWh]	2500 Full Load Hours [EUR/MWh]
Investment (12 year annuity at 4%)	40 to 50	30 to 40
Operation and maintenance including major overhauls	12	12
Other operational expenses	8	8
Total	**60 to 70**	**50 to 60**

Currently, the costs of wind energy are slightly higher than the feed-in tariff for electricity produced from conventional fossil fuels or by nuclear power plants.

However, most European countries have incentives to stimulate the production of renewable energy, including wind energy. Although each country or state applies its own rules, common features are:

- Tax benefits on investments in new renewable energy (wind) assets;
- Grants on installing new renewable energy (wind) assets;
- Lower interest rates from green funds for financing renewable energy (wind) assets;
- Incentives on the production (kWh) of renewable energy (wind).

As a consequence of one or more incentives, investments in wind energy can be profitable. In the past, tax benefits up to 50% of the investment costs were not uncommon. In the case of a feed-in tariff, including incentives, of 80 to 100 e/MWh, the cost recovery period is between 6 (> 2700 full load hours) and 10 (> 1900 full load hours) years.

Site Selection Wind Energy

Next to issues such as sufficient space for an envisaged wind farm site, accessibility by heavy equipment such as cranes, limited nuisance to the neighborhood and the presence of a medium voltage substation with sufficient capacity, the most important issue is the presence of sufficient wind.

As a first guide, investors and developers may use the European Wind Atlas [2] to estimate the long-term wind speed. A second source is wind data from meteorological stations located up to 30 to 40 km from the site. In case of wind farms installed in the surrounding, say less than 5 to 10 km, these production data may also serve as a good starting point.

More in detail the site wind speed and subsequently the envisaged wind farm output can be predicted with the WAsP software tool [3]. WAsP needs the long-term wind speed distribution of at least three surrounding meteorological stations. The accuracy of the results increases when the met stations are located close to the potential site. Subsequently the envisaged site and its surroundings, more precisely the surface roughness, are modeled as accurately as possible. The output is the long-term wind climate at site.

In case of doubt, certainly for complex terrain like hilly and mountainous areas, additional wind measurements are required. The measurement period has to be at least a year and if possible extended to two years.

Project Risks

The main project risk is that the long-term wind climate at the potential site is lower than anticipated during the feasibility phase. As a result of the cubic law relation between the wind speed and the power, a relatively small decrease in the long-term wind speed may have a large effect on the energy output. A significant lower energy yield, i.e., more than 10 to 15%, may result in cost recovery times of more than 12 to 15 years instead of less than 10 as estimated. The result is a loss-making project.

It is therefore advisable to use a somewhat lower mean wind speed in the financial and economic calculations. Instead of using a wind speed with a 50% probability of being exceeded, use a lower value with an 80 or 90% probability of being exceeded. By doing this, in 8 to 9 out of 10 years one is assured of a wind speed, and therefore an energy output, higher than estimated.

The following items have to be considered when building a wind turbine:

- There must be sufficient space and plenty of wind. Nearby the applicable area many deflections can occur. These deflections can, for example, be caused by hill slopes or obstacles.

- The territory must have a permit to operate a wind farm. In practice, this means that mostly territories that are marked as industrial areas are to be considered. Otherwise it is necessary to change the local zoning scheme.
- The territory must be accessible. During the erection of the wind turbine it is necessary for large hoisting cranes to be able to reach the construction site.
- It has to be possible to connect the wind turbine to the electrical grid. The voltage level can be from 10 to 30 kV when connected to the local distribution grid. In the case of a wind farm, where the generated power is much larger, it is necessary to connect at the voltage level of a transmission grid.

Wind Turbine Power Control

The previous chapter demonstrated that the power increases with wind speed according to a cubic law. Most wind turbines reach maximum power, also called the rated or nominal power, at wind speeds between 12 and 14 m/s. At higher wind speeds, the power has to be kept constant in order not to overload the wind turbine structure or the electrical connection.

Wind turbine technology applies the following methods to control the power above the rated wind speed (see also Figure 2):

1. **Stall Controlled Rotors**

   The rotor is kept at a constant speed and the mostly asynchronous generator is connected to the 50 or 60 Hz public grid without the use of a converter or other power electronics. Power control is based on the aerodynamic principle that if the flow angle-of-attack reaches a certain limit (stall point), the lifting force and subsequently the rotor torque stabilizes or even decreases in magnitude. The main advantage of this concept is its simplicity; no mechanical or electronic systems are required to limit the power because this is a completely passive system. In early days of modern wind technology, the 80s and 90s of the last century, stalling was the most widely used power control system. The Danish wind turbine manufacturers in particular gained extensive experience in this control principle. Currently stall control is not often applied any longer. The main reason is that when stall control is applied to a wind turbine greater than 1 to 1.5 MW, this may lead to resonance problems in the rotor blades and drive train. Another disadvantage is the relatively poor power quality obtained from stall wind turbines.

2. **Variable Speed Rotors**

   Although this concept was already known and also applied on limited scale in the 80s and 90s, this control mechanism has been developed further and used widely since the end of the 90s. The rotor speed is variable and increases in proportion to the wind speed. At the rotor speed producing the nominal power; the power is kept constant by pitching the blades towards the wind. By pitching the blades into the wind direction, the angle-of-attack is lowered and the lifting force and rotor torque are reduced. The synchronous generator is connected to the grid using a converter or other power electronics that can deal with alternating frequencies. The advantage of this control mechanism is that it can be applied to MW wind turbines without introducing undesirable mechanical resonances. Applying blade pitch together with up-to-date control techniques may lower design loads and may serve as a good starting point for further up-scaling activities. Last but not least, modern, IGBT or IGCT-based, converter technology improves the quality of the generated power.

3. **Intermediate Power Control Solutions**

   In the past two decades, several power control methods have been introduced that are based in one way or another on the above-mentioned control mechanism. A power control mechanism applied in the 90s by a limited number of manufacturers is the so-called "active stall" control. This method combines stall control including constant rotor speed with blade pitch to optimize the stall characteristics. Another variation is the combination of stall/constant speed and power electronics to optimize the power quality.

(More Details are Given in Reference [1])

Wind Power Applications and Opportunities in Different Sectors

The owner or operator of a wind turbine mostly sells the electricity produced to the utility company. Various systems of price rates have been determined in the EU to structure the trade in electrical power.

Owners or operators can be:

- Initiatives from private persons or companies. They finance wind projects with own money or loan capital. Many taxation regulations are applicable for companies.
- Cooperatives; private persons create a legal structure to set up a wind turbine or wind farm together. Shareholders participate in profits according to the efficiency of the turbine.
- Utility companies; as turbines grew in capacity, more utilities became interested in the wind industry. The companies are particularly interested in large wind farms and will probably participate in the development of new offshore wind farms.

Current Status of Wind Power

The manufacture of commercial wind turbines started in the 1980s, with Danish technology leading the way. From units of 40-60 kW with rotor diameters of around 10 m, wind turbine generators have increased in capacities over 5 MW and rotor diameters of 120 m and above.

Continual improvements are being made in the ability of wind turbines to capture as much energy as possible from the wind. One result is that employment in the European wind energy sector has been growing rapidly. In Denmark, for example, the number of people employed both directly and indirectly in wind turbine manufacture increased from about 2,900 in 1991 to 21,000 by 2002.

Estimations based up on the EWEA scenario "Wind Force 12", employment in Europe could reach almost 200,000 by 2020, with double that number for global employment.

Other facts of wind power worldwide and in Europe are:

- By the end of 2005, worldwide 60,000 MW wind power capacity had been installed.
- Over the last few years, the worldwide yearly increase has been approximately 25%, worldwide in 2004 7,500 MW and in 2005 11,600 MW wind power capacity has been installed.

- The majority, 60 to 704.
- It is estimated that 15,000 MW of wind power will be installed worldwide in 2006.
- Outside Europe, most wind power is installed in USA and the last years China and India are rapidly increasing wind power capacity.
- Wind energy has grown most consistently in Europe, with capacity multiplied by 27 times over the decade between 1992 and 2002.
- The leading nations in wind energy are Germany, Spain, Denmark and the Netherlands, who account for 84% of the total European wind capacity. Emerging markets include Austria, Italy, Portugal, Sweden and the UK. All ten Member States that joined the EU in May 2004 have also adopted targets for the level of renewable energy they are expected to achieve.
- In Germany last year, the wind industry turnover was €4.2 billion.

Figure 3: Map of Europe with Installed Amount of Wind Power per EU-Member in MW

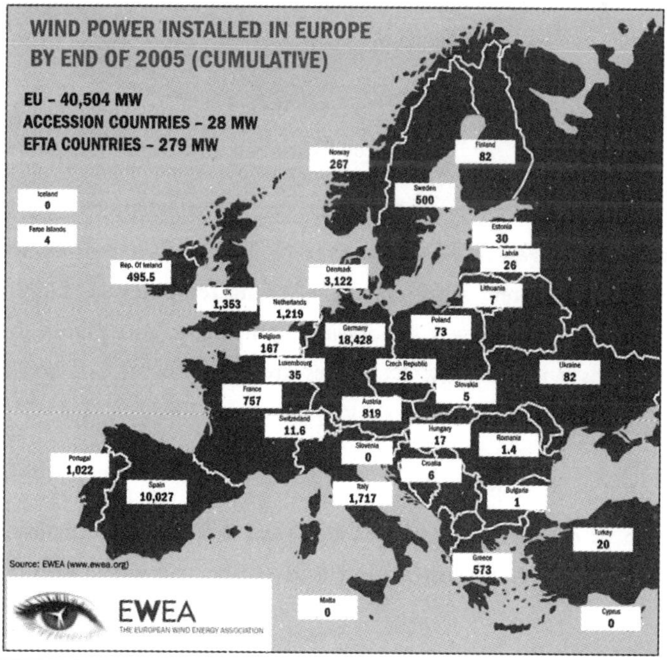

Wind Turbine Manufacturers

Worldwide, the market for wind turbine manufacturing and installation is dominated by a limited number of manufacturers. The leading ten manufacturers have a worldwide market share of 95% in wind turbine production and installation, see Figure 4.

Figure 4: Worldwide Market Share of Wind Turbine Manufacturers (Production and Installation)

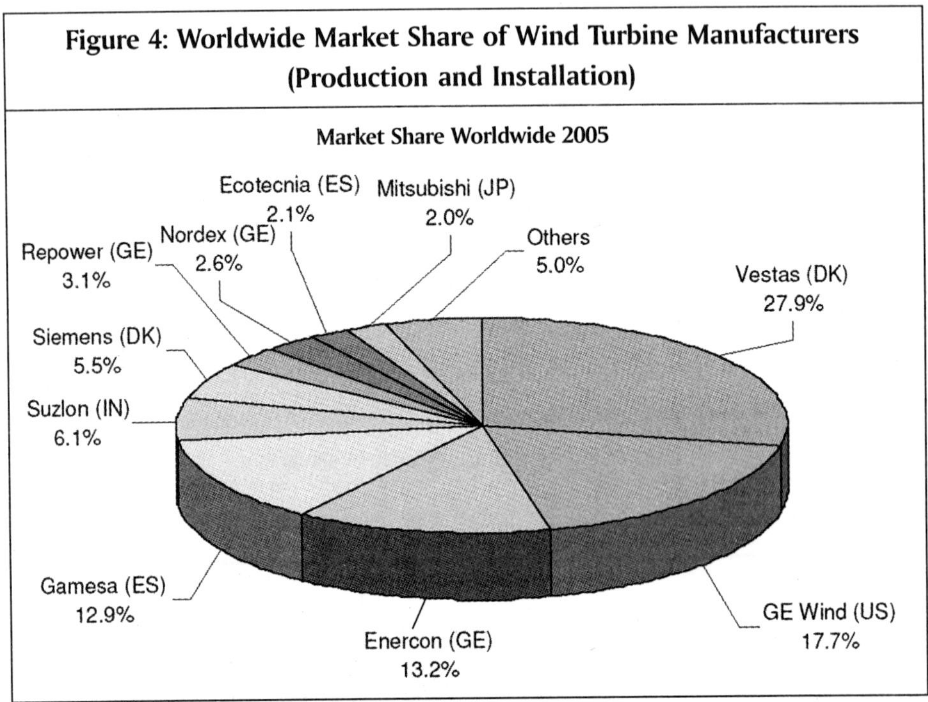

Market leader is Vestas Wind Systems (VWS) with a market share of almost 30%. Since approximately 2000 there is growing tendency for large multinational operating companies in power generation equipment and plants to expand their activities towards wind energy. Examples are Mitsubishi, GE and Siemens with the takeover of Bonus from Denmark.

Trends

In perception of economical and technological aspects, three major trends can be recognized in recent years concerning grid connected wind turbines:

- **Turbines have grown larger and taller**

  The average capacity of turbines installed in Germany and Denmark increased from approximately 200 kW in 1990 to almost 1.5 MW during 2002. Turbines in the 1.5 to 2.5 MW range have more than doubled their share of the global market from 16.9% in 2001 to 35.3.

- **Investment costs have decreased**

  The average cost per kW of installed wind power capacity currently varies from 900 €/kW to 1200 €/kW. The turbine itself comprises about 80% of this total cost. Foundations, electrical installation and grid connection mainly account for the remainder of the cost. Other costs are land, road construction, consultancy and financing costs.

- **Turbine efficiency has increased**

  A mixture of taller turbines, improved components and better siting has resulted in an overall efficiency increase of 2-3.

In addition to the previously mentioned trends, there is also the fact that offshore wind farms have become larger in number and size. In the beginning, offshore turbines were "sea-adjusted" versions of land-based technology, with extra protection against sea salt incursion. Present generations include more substantial changes; such as higher peripheral rotor speeds and built-in handling equipment for maintenance work. The turbines must be firmly positioned on the sea bed based on a precise design. Many kilometers of cables have to be laid both between individual turbines and back to shore to feed the generated power into the grid. To ensure a high reliability of wind turbines, it is of great importance to perform effective maintenance on turbines. This requires service vessels that can transport maintenance crew in extreme weather to the turbine platforms.

By the end of 2003, a total of almost 600 MW of offshore wind farms had been constructed around Europe in the coastal waters of Denmark, Sweden, the Netherlands and the UK.

Wind Turbine Technology

The technology of modern wind turbines has developed rapidly during the last two decennia. The basic principle of a wind turbine has remained almost unchanged and consists of two conversion processes. These processes are performed by its main components:

- The rotor that extracts kinetic energy from the wind and converts it into generator torque
- The generator that converts this torque into electricity and feeds it into the grid.

Although this sounds rather straightforward, a wind turbine is a complex system in which knowledge from the areas of aerodynamics and mechanical, electrical, and control engineering is applied.

Rotor and Wings

A modern wind turbine consists of two, preferably three wings or blades. The blades are made of polyester strengthened with glass fiber or carbon fibers. On a commercial basis, wings are available from 1 meter up to 100 meter and more. The wings are mounted on a steel construction, called the hub. As mentioned, some wings are adjustable by pitch control.

Nacelle

The nacelle can be considered to be the machine room of the turbine. This housing is constructed in such a way that it can revolve on its (steel) tower to face the rotor into the right wind direction. This facing into the wind is controlled by a fully automatic system and is set by a pennant on the turbine housing. The machine room is accessible from the tower and contains all the main components such as the main shaft with bearing, gearbox, generator, brakes and revolving system. The main shaft transfers the rotor torque to the gearbox.

Gearbox

A gearbox increases the number of revolutions from the shaft to the desired number of revolutions of the generator. A turbine with a capacity of 1.0 MW and a rotor diameter of 52 m turns at about 20 revolutions per minute and the generator at 1500 rpm. The necessary transmission rate will be: 1500 divided by 20 is 75.

Figure 5: Cross-Section of a Wind Turbine Nacelle

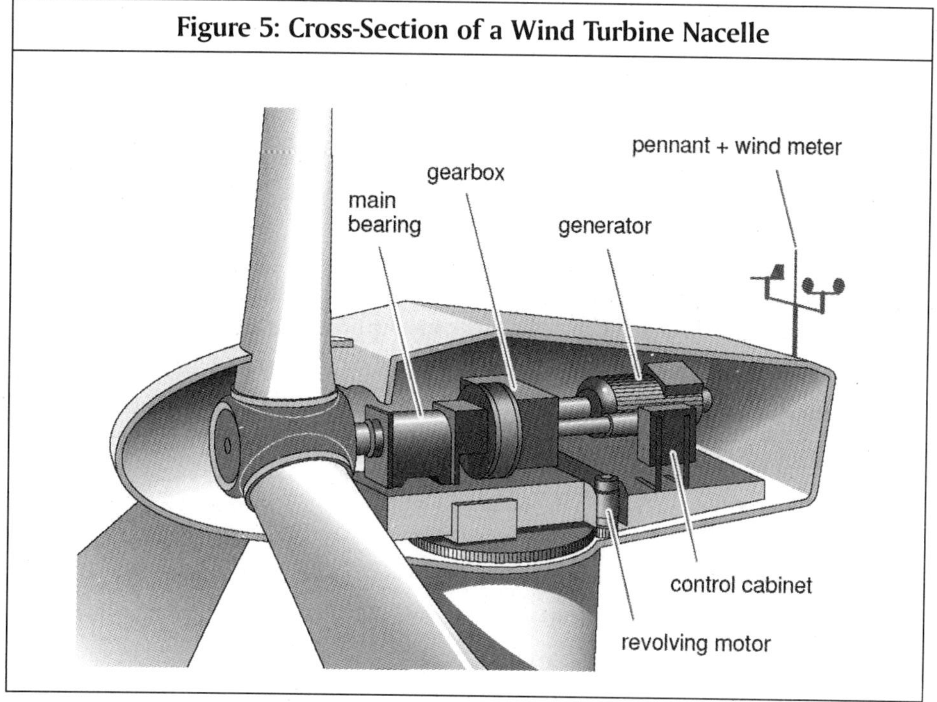

Generator

Currently, there are three main wind turbine types available. The main differences between these concepts concern the generator and the way in which the aerodynamic efficiency of the rotor is limited during wind speeds above the nominal value in order to prevent overloading. As for the generator, nearly all wind turbines currently installed use either one of the following systems (see Figure 6):

- Squirrel-cage induction generator
- Double fed (wound rotor) induction generator
- Direct-drive synchronous generator.

An asynchronous squirrel cage generator is a wind turbine with the first generating system. Because of the great difference between the rotation speed of the turbine and the generator, a gearbox is used to couple them. The stator windings are connected to the grid. This concept is called a constant speed wind turbine, although the squirrel cage induction generator allows small variations in rotor speed (approximately 1%). A squirrel cage generator does consume reactive power from the grid. This is not a desirable situation, especially in a weak grid. For this reason, the generator's need for reactive power is compensated for with capacitors.

The other two generating systems allow for about a factor of 2 between the minimum and maximum rotor speed. These different speed levels are intercepted by the decoupling of grid frequency and rotor frequency. For this, decoupling power electronics is used.

Figure 6: From Left to Right, Commonly Applied Generating Systems in Wind Turbines: Squirrel Cage Induction Generator, Double-Fed (Wound Rotor) Induction Generator, Direct-Drive Synchronous Generator

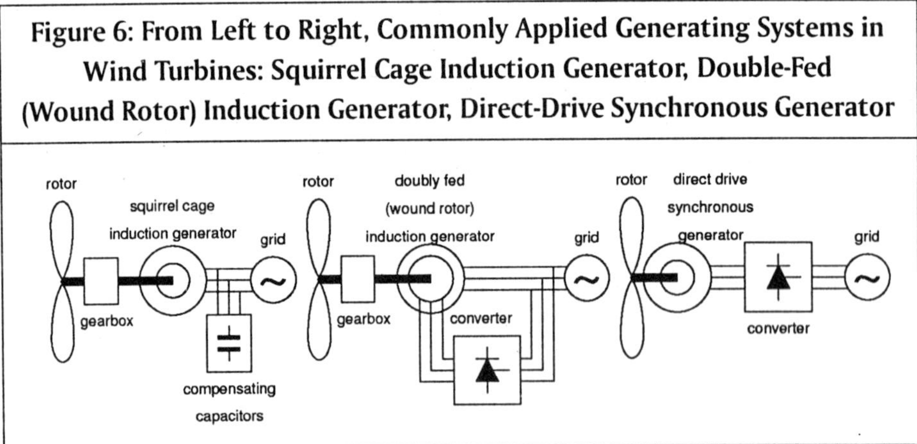

The first variable speed concept is based on the double-fed induction generator. Through the power electronics, a current is injected in the rotor windings of the generator. The stator windings of the generator are directly connected to the grid. The frequency of the current injected into the rotor windings is variable so electrical and mechanical frequencies are decoupled. By doing so, operation with variable speeds is possible. A gearbox adapts the two different speeds of rotor and generator.

A direct drive synchronous generator is used in the second variable speed concept. The additive "direct drive" refers to the fact that these turbines don't have a gearbox. Generator and grid are fully decoupled by power electronics. In this configuration, variable speed operation is also possible. For this concept, some manufacturers use special low revolution generators. Generators with low speeds are recognizable by their relatively large diameters that are positioned close to the turbine rotor.

As can be concluded from this description, there is a fundamental difference between conventional thermal or nuclear power generation on the one hand and wind power on the other, namely: in wind turbines, generating systems different from the synchronous generator used in conventional power plants are used.

Braking System

Wind turbines are equipped with a safety system of a high degree. An aerodynamic braking system is part of this. In case of emergencies, or for parking the turbine for maintenance, a (disk) brake is usually fitted.

Control System

Wind turbines are high-tech machines. After commissioning, a turbine is controlled fully automatically by an internal computer system. Information on the status of the turbine can be retrieved remotely by owner or manufacturer by telecom transmission (e.g. modem).

2.2 Future Developments

At the moment, wind turbines with proven technology are available in the range of 1.5 to 3.0 MW. Especially in Western Europe private wind farm developers and utilities focus on wind turbines in the power range of 2.0 to 3.0 MW. All leading manufacturers have one of more wind turbines in the MW+ segment.

In some countries, e.g. Southern Europe, Asia and Latin America, with less developed transport infrastructure or dominated by mountainous areas, wind turbines having more limited dimensions are more appropriate. This is the reason why wind turbines in the power range between 0.8 and 1.3 MW are most in demand worldwide.

Wind turbine prototypes are available in the power range 5 to 6 MW, these turbines are characterized by a shaft height of 120 meter or higher and rotor diameters of more than 110 meter. These wind turbines will become commercially available from 2006. Besides the still high costs of these 5+ MW turbines per installed MW, the main problem is the weight and outer dimensions of the components. Components are of sizes that are hard to transport over the Western

European road infrastructure. Some manufacturers solve this problem by offering these turbines only for offshore sites or sites accessible by waterways. Other manufacturers manage this logistic problem at least partly by building and installing the towers from prefab concrete parts or in-situ concrete tower structure instead of tubular steel segments.

In wind turbine technology, the following developments are under preparation on the drawing boards or are anticipated by experts:

- The market share of variable speed rotor technology, including modern power electronics, will increase further.
- Also in the segment of MW+ wind turbines, the gearbox is one of the weakest links, requiring frequent maintenance, and refurbishment or replacement is expensive. A select number of manufacturers offer gearless wind turbines. These wind turbines are characterized by large, with diameters up to 5 meter, multi-pole synchronous generators. Hybrid designs are also obtainable, e.g., a one-stage step-up gearbox followed by a less massive multi-pole generator. For the coming 5 to 10 years it is anticipated that these different concepts will be developed side by side.
- The development of MW+ wind turbines will focus on saving weight and limiting dimensions in order to simplify road transport and limit the required capacity of building cranes on site. Ways to achieve this are the optimization of the control strategy leading to less loaded and so less massive components. Another method is to integrate more functionality in a component or system leading to less or more compact parts.
- Currently offshore wind turbines are equal to or derived from onshore turbines. In the near future the development of offshore wind turbines will deviate from onshore wind turbines. Offshore wind will focus on reliability, remote control and high installed power per wind turbine (up to or more than 10 MW). Onshore wind will concentrate on low and acceptable nuisance (i.e. noise) for the neighborhood, high efficiency, easy and low cost transport to the site, installation by fast and on short notice available building cranes, and limited installed power (up to 6 to 8 MW).

Costs and Benefits

Costs of Wind Power

Cost prices of wind power depend to a considerable degree on the location of the turbine. The wind speed and costs for the connection to the grid can vary per location. For commercial use (budget and depreciation over ten years), cost prices vary from 5c€/kWh on good windy areas up to 8c€/kWh inland. In comparison: the cost price of electricity generated by fossil fuels is approximately 4c€/kWh. Payment for the delivered energy consists of avoided fuel costs, partly ecotax (grants for green energy) and a part that is determined by the market for renewable power. This cost calculation is based on the following assumptions:

- A new medium-sized wind turbine of 850 – 2500 kW capacity
- O&M costs averaging 1.2c€/kWh over a lifetime of 20 years. The aggregated operational expenses (land rent, taxes, insurance, daily operation, O&M) is approximately 2c€/kWh.

During the last twenty years, the investment costs per kW installed wind power has dropped. These average costs for a wind turbine project range from €900 up to €1200 per kW installed power. Since the start of the present development in the end of 1970s, generating costs have dropped by about 80%. Expectations tell us that this trend will continue with a reduction of a few percent each year.

The other principal cost element in generating electricity from wind power is operation and maintenance (O&M). Obviously, there are no fuel costs. O&M costs include regular maintenance, repairs, insurance, spare parts and administration. Because not many machines are more than 20 years old, data is not always available, or comparable. For a new machine, O&M costs might have an average share over the lifetime of the turbine of about 20-25% of the total amortized cost per kWh produced. Manufacturers are aiming to reduce these costs significantly through the development of new turbine designs requiring fewer regular service visits and therefore reduced downtime. The trend towards larger wind turbines also reduces O&M costs per kWh produced.

Except for investment and O&M costs, costs for the following items also have to be taken into account:

- Project development
- Preparation of the building site
- Bedplate of the turbine
- Connection to the grid
- Real estate taxes.

Benefits of Wind Power

The owner of a wind turbine sells his generated electricity to a utility company. The value of wind power, as seen by the utility, is to be determined by the costs as caused by the replacement of coal or gas. The average price of this grey power has gone up to approximately 2 to 3c€/kWh. If the owner were only to be compensated for this part, a wind turbine could not be utilized sufficiently when the present costs are taken into account.

Utility companies also pay for a guaranteed supply of power. "Back up" power is not needed if the power supplied as a high level of availability. Statistics have shown that wind power can, in slow wind speeds, represent approximately 25% of the guaranteed power.

Future Costs

Can wind power be competitive with conventional existing power plants at this moment? In this comparison, wind power does not have an advantage because of existing plants that are partly written off. Decisive is, how electricity made by wind will act in ten years in relation to electricity produced by then newly built conventional plants that are powered by fossil fuels; of which all the exhausted gasses need to be cleaned and probably all emitted CO_2 has to be stored. Because sources of fuels are running out, there's a realistic chance that the prices of fossil fuels will remain high. On the other hand, the costs of wind power are expected to continue dropping.

If wind power works out in a positive way for the next ten years, it has a good chance of becoming a serious competitor to conventional energy sources.

Taxes and Incentives

In most European countries, wind power has at this moment no chance of surviving without an incentive contribution from EU governments.

A substantial reason for providing an incentive is that wind power, as (almost) clean energy source has almost no external costs. These costs have been recently determined as part of a study made by the EU. The European Union states that "external costs are incurred when the social or economic activities of one group influence another group, and when this influence is not entirely being compensated or taken into account". For example, a conventional power plant produces SO_2. This gas causes a shortness of breath in people who are asthmatic and damage to building materials. However, the owner of the plant does not pay for the extra health care or the repair of buildings. The owner shifts these costs to others. The EU could introduce an ecotax for this damage. As a consequence of this, the energy per kWh would be 2 up to 7c€ more expensive.

The other way around is to encourage clean energy sources by incentives, so social and environmental costs are avoided. These subsidiaries are allowed, if not encouraged by the EU. In some European countries for example, wind power is rewarded with subsidy rates of approximately 8 or 9c€/kWh depending whether it is located on land or in the sea.

Policy and Regulation

Relevant Regulations, EU Policies and Directives Concerning Wind Power

There are major disadvantages in being dependent on oil imports and fluctuating oil prices. Sources of fossil fuels are running out. The environment is also becoming a major issue in terms of CO_2 emissions or how to store nuclear waste.

In the background, many industrialized countries are exerting a major effort to develop renewable energy sources; especially solar, biomass, hydro and wind power. Shell expects that one third of all energy demand worldwide will be provided by renewable sources in the year 2050. The ambitions to apply wind power are high. Some countries within the European Union have set individual targets to have (for example) 9% of all electricity be generated by renewable power by the year 2010. Half of this percentage might be wind.

These ambitions are very modest compared with the targets that have been set by the European Union. Most countries of the EU now already have a higher share of renewable power generation by the availability of hydropower and a higher use of biomass and wind power. In the former EU (15 countries) and by the year 2010, 22% of all generated electricity must be produced by renewable energy sources. The Union is constantly encouraging all former and new European members to achieve this goal.

Local Effects of Wind Power

When wind power is used, this can cause local inconvenience. Detailed planning can limit the impact of wind power consequences to an acceptable level.

Birds

Birds can run into the rotor blades of a turbine or get caught by the turbulence behind the rotor. Inquiry has shown that the risks for collisions are relatively small. The estimated number of "collision-victims", at an installed power of 1000 MW is approximately 21,000. On an annual basis this seems to be a high number, but this is a small amount when compared to the number of birds that get killed by traffic (2 million) or that die because of power lines (1 million).

Most wind turbine casualties are caused at night, during twilight or in bad weather situations. Birds know their forage and resting grounds very well; they avoid wind turbines. When installing turbines, the breeding and foraging areas of the birds have to be considered well.

Fish

Wind turbine (farms) at sea also have positive effects. Over-fishing is a well-known problem; the stocks of many species of fish are threatened. The sailing of ships and therefore fishing is prohibited in and around wind farms. Sea biologists expect that these areas will develop into a breeding ground of several species of fish, and overall this will have a positive effect on fish stocks. Recent research near wind farms confirms this effect.

Noise

Wind turbines produce noise. The rotor makes a zooming sound and the generator and gearbox can also be heard. Carefully designed rotor blades, a limited revolution speed, and effective sound insulation of the gearbox and generator limit the noise emission. By maintaining a sufficient distance from residential or other sound-sensitive areas, noise pollution can also be avoided.

Shadow

Sunshine creates moving shadows when the rotor of a turbine rotates. In winter, when the sun is low, the shadow can be annoying when it falls into a window. Giving wind turbines an appropriate orientation towards houses is sufficient to prevent this problem. If on a yearly basis a small number of hours give inconvenience, the turbine can be stopped at these moments without any particular loss of energy.

Blending into the Landscape

Wind turbines are striking structures in the landscape. They can be made to blend in by, for example, by arranging them in lines along a dike or waterway. In doing so, the lines of the landscape are taken into account. Research has shown that positioning wind turbines in clusters is more accepted when it is clear to neighboring people that in this situation a great yield can be generated. Whether the lining-up of several turbines is liked or not is and always will be a matter of taste. More important is the relationship between the altitude of the shaft and the diameter of the rotor. Another significant item is the size of the rotor. Rotors that have larger diameters rotate slower and because of this they are quieter.

Summary

For thousands of years, wind power has been used for various purposes. Mainly since the oil crisis a substantial development has taken place and impressive wind farm projects have been realized.

Wind technology is still developing. Turbines are becoming more efficient, power rates are increasing and intelligent power electronics are being introduced. In the meantime, impressive wind farms are rising out of the sea.

Not only environmental issues make wind turbines more often seen on the horizon. The continuously lowering of the investment and maintenance costs of wind turbines make this technology interesting for investors and wind farm developers.

(Ton van de Wekken, and Fred Wien are from KEMA Consulting, founded in 1927, it is a commercial enterprise, specializing in high-grade business and technical consultancy, inspections and measurements, testing and certification. Authors can be reached at Ton.vanderWekken@kema.com ; fred.wien@kema.com).

References

[1] Thomas Ackermann (ed.), *Wind Power in Power Systems,* John Wiley & Sons, Ltd, 2005, ISBN 0-470-85508-8.

[2] Troen, I. and E.L. Petersem, *European Wind Atlas,* RisϕNational Laboratory, Roskilde, Denmark, ISBN 87-550-1482-8.

[3] WAsP (Wind Atlas Analysis and Application Program), software version 8, RisϕNational Laboratory, Roskilde, Denmark.

[4] Jos Beurskens, Gijs van Kuik, Alles in de wind, answers and questions concerning wind power, October 2004.

[5] Wind power technology, operation, commercial developments, projects, grid distribution, EWEA, December 2004.

[6] Wind power economics, wind energy costs, investment factors, EWEA, December 2004.

[7] "The current status of the wind industry, industry overview, market data, employment, policy", EWEA, December 2004.

[8] Windenergie winstgevend, Ministry of the Flemish Community, Department of renewable sources and energy, 1998.

3

Whose Wind?

Prospects for Cooperative and Community Wind Development on the US Upper Great Plains

Bart Finzel and Arne Kildegaard

The Upper Great Plains of the US have a tremendous wind resource which is expected to be developed for electricity over the next decade. The region is also notable for the strong foothold of the cooperative movement, dating to the 19th century. While other research has emphasized the relative economic advantages of locally-owned wind to rural communities, here we analyze the potential for the anticipated growth in this industry to be organized cooperatively – or at least locally. Rural Electric Cooperatives (RECs) are natural candidates to spearhead the development, but have thus far largely viewed as a threat rather than as an opportunity. The "community-wind" movement represents an alternative form of local ownership. While independent community wind developers lack some of the natural assets of the RECs, they have the competitive advantage of being able to incorporate and finance themselves in ways precisely tailored to capture the greatest tax incentives. The social costs, however, are almost certainly higher with community wind than with REC wind.

I. Introduction

The Upper Midwest of the United States (chiefly the Dakotas, Iowa, Minnesota, the Upper Peninsula of Michigan, and Wisconsin) is a traditional center of strength and innovation in rural cooperative enterprises. The region includes large agricultural cooperatives (nine of the ten largest in the US) and a robust, eclectic group of smaller cooperatives. In Minnesota, for example, there were 841 cooperatives and 185 credit unions in 2001, numbers that make the state the nation's leader.[1] These cooperatives provide products and services in a variety of industries, including agricultural processing and marketing, health care, housing, food, rural fire protection, funeral homes, and educational services.[2] Included in this number are 43 rural electric cooperatives (RECs) that distribute power in rural areas, and 3 generation and transmission (G&T) cooperatives that provide their services to the RECs. The creation of these cooperatives involved a determined effort on the part of rural people to organize collectively to provide a service for themselves and their neighbors and to create economic development opportunities for their communities.

The Upper Midwest is also the region in the US with the greatest potential for electrical generation from on-shore wind.[3] Wind-power constitutes a small but dynamic segment of the electricity industry in the United States, with growth of fully 45% in 2006 alone.[4] Tremendous growth is projected over the next two decades, in part as a result of the ever improving economics of wind generation, and in part as a result of legislative mandates (renewable electricity standards) that stipulate escalating percentages of renewable electricity.

1 See Folsom (2003) for greater detail. Minnesota is also home to the so-called "Minnesota Model," a reference to that state's cooperative law passed in 2003. The law allows for limited inclusion of passive investors, as well as patron/customers, in order to address some of the equity challenges that has led to increasing preference for limited liability company (LLC) forms of ownership. Whether the law goes too far in this direction and threatens to undermine democratic decision-making in the cooperatives is an open question. See O'Brien (2005).

2 Spear (2000) identifies several potential advantages to the cooperative organization, several of which are relevant in explaining the origins and persistence of the movement in the Upper Midwest: cooperatives are adept at providing quasi-public goods when markets fail; cooperatives are effective agents to assist weaker actors when combating excessive market power; cooperatives have a "trust" advantage, being perceived as less likely to engage in opportunistic behavior and exploit consumers; cooperatives may build on solidarity within communities to enhance their social capital and improve economic performance; and, lastly, cooperatives generate positive externalities through community empowerment.

3 See for example the National Renewable Energy Laboratory (NREL) windmaps, available on-line through *www.eere.energy.gov/windandhydro/windpoweringamerica/wind_maps.asp.*

4 10,492 MW of installed capacity were in place as of September, 2006, and several thousand additional MW are either in the construction or planning stage according to American Wind Energy Association's website, (last accessed 1/08/07): *http://www.awea.org/projects/*. Nevertheless, actual generation from wind constituted only about .6% of total generation in 2006 according to the US Energy Information Administration (see *http://www.eia.doe.gov/cneaf/electricity/epm/tablees1b.html* last accessed 1/08/07).

All of these trends make this an opportune moment to assess the ownership and control structures under which the industry might develop in the high-growth years ahead. Several studies have shown that the regional economic impact of "corporate" wind development is only a fraction of what could be achieved with local ownership.[5] In light of the region's wind resource endowment, its historical tie to the cooperative movement, and the magnitude of what is at stake for some of the most impoverished rural counties in the country, this paper examines the potential for the anticipated future growth in the wind-generated electricity industry to be organized cooperatively, or at least owned locally.

To date, neither the distribution cooperatives nor the G&Ts have played a major role in the nascent wind industry. The wind power that has been so far developed is overwhelmingly of the "corporate wind" variety—usually wind farms developed by an investor-owned utility, frequently on a scale exceeding 100 MW. A growing "community wind" movement has sought to shape public policy in ways more favorable to community ownership and cooperative management principles, but ironically, the RECs which seem most logically suited to the task have not taken the lead. In fact, the RECs have tended to perceive wind power—as well as public policies aimed at its promotion—as a threat rather than as an opportunity.

The paper begins by briefly describing the emergence of consumer and producer cooperatives on the Upper Great Plains and the motivation for their formation. We next discuss the RECs in their historical origins and purposes and current operations. We examine the apparent advantages these structures have relative to other community wind organizations for the development of wind in the future, and we explore the challenges they face—organizational, regulatory, and contractual—that may make it cumbersome for them to evolve along these lines. Finally, we discuss the community wind movement more broadly, in its various incarnations, including some of the particular niches it has found and successfully exploited. We conclude with policy recommendations of interest both to wind power and rural development advocates.

II. A Cooperative Context

The north-central region of the US has historically been home to the most significant cooperative movement in North America. Both producer and consumer cooperatives

---

5 See for example US General Accounting Office (2004).

were common mechanisms for mutual aid and self-help during the period of rapid European settlement in the region in the late nineteenth century.[6]

Kercher, *et. al.,* (1941) traces the origins of most modern cooperatives in the region to the wave of Finnish immigrants into Northern Michigan, Minnesota, and Wisconsin in the early 20th century. The remoteness of the region and the isolation of the settlements often plagued these communities with uncompetitive markets for provisions. Cooperatives were formed as a collective instrument of material self-help, then, but not without an ideological dimension "to usher in a more equitable social order" (Kercher, p. 16) as well. The movement grew quickly throughout the north-central states. In the mid-1930s, 70% of the cooperatives identified by the US Bureau of Labor Statistics were located in the region.

While nearly all of these enterprises were/are organized primarily as consumer cooperatives roughly along Rochdale principles,[7] three broad types emerged: consumer's retail distributive societies, farmer's purchasing societies, and special service and supply cooperatives.

The retail distributive societies simply bought goods collectively in order to achieve the lowest possible prices, consistent with quality. The objectives of the farmer's purchasing societies and the special service and supply cooperatives, however, are not so easily characterized. In nearly all cases, but with varying degrees of emphasis, these have historically served the dual consumer and producer interests of their members. Many farmer cooperatives serve the producer interests of their rural members by purchasing commodities (feed, machinery) used directly in production, while an overlapping subset has provided members a mechanism to market or distribute their commodity production to others.

In the case of special service and supply cooperatives, the cooperatives are ostensibly organized by the consumers of the service. Water, telephone, health care, insurance protection, credit, and electric services were and are provided by numerous consumer

---

6 The "cooperative coopers" of Minneapolis is perhaps, the most notable attempt to create producer cooperatives in the region. In the 1880s, barrel cooperatives employed hundreds and inspired a cooperative store, a painter's cooperative, a cigar making cooperative, and a cooperative laundry and led to the formation of a short-lived cooperative colony. See Leiken (1999).

7 These include open and voluntary membership, democratic control, and limited interest rate on capital with return of surplus earnings to patrons in proportion to patronage.

owned cooperatives or mutual aid entities in the region. However, many of these organizations have farmer and small business members and, of course, render services that are used directly in productive enterprises. This is especially true of entities whose membership base is rural people, nearly all of whom rely on the service to enable production and increase income as well as to "buy for less". As with farmers' organizations, identifying these solely as consumer-based entities ignores the dual consumer and producer interests of members.

III. The Rural Electric Cooperatives

The first REC is believed to have been established in Granite Falls, Minnesota in 1914.[8] By 1923, 31 electric cooperatives had been incorporated in nine states.[9] While the movement was well underway already, it was given an enormous boost when President Roosevelt established the Rural Electric Administration (REA) by executive order in 1935. Congress passed the Rural Electrification Act in 1936, establishing a loan program within the REA to create a national rural electrification program.

In 1935, fewer than ten percent of farmers in the United States had access to electric power. In that same year, a special committee of investor-owned utility (IOU) executives gave a report to the REA which concluded "there are very few farms requiring electricity for major farm operations that are not now served."[10] The statement of course was mistaken: over the next 20 years rural electrification would reach more than 90% of US farms, and contribute significantly to a surge in agricultural productivity.[11] But the statement also betrays a certain disdain for serving rural areas on the part of the for-profit industry at that time.

News of the success of the electric cooperatives spread. The Farm Bureau and other major farm organizations began, in late 1935, to encourage farmers to form cooperatives to obtain electrical power. The REA also began actively to encourage and assist the nascent electric cooperatives.[12] In the void created by IOU indifference and apathy, it had became clear that rural electrification would best proceed working outside of the for-profit sector. The IOUs were not particularly distressed by this competition, but

8 Voorhis (1961).

9 Ellis, p. 34.

10 Quoted in Ellis, p. 39.

11 Gardner (2002).

12 Seven of the first ten REA loans went to cooperatives.

neither did they make life easy for the cooperatives.[13] Voorhis (1961, p. 57) explains how the organizers of these first cooperatives built upon a critical but intangible asset: "[O]rganized into these cooperatives...was a common need of a group of neighbors. In fact, these electric cooperatives—as well as almost any kind of cooperative that is well conceived—consist basically of the organization of a common need of a group of people." Organizing around a shared need built social capital and lowered development costs by calling forth free-will offerings of labor (for digging holes, setting poles, and other cooperative tasks) and millions of miles of rights-of-way for the power lines themselves. The effects on production were felt everywhere, but were perhaps most significant in large tracts of the High Plains, where cheap electricity enabled exploitation of groundwater reservoirs. After several historical failed attempts to farm the High Plains, with their variable and uncertain precipitation, decades of successful High Plains farming now owe directly to the introduction of electrical irrigation, made possible by the formation of the rural electric cooperatives in the 1930s.[14] In many of the RECs overlying the aquifer, the electrical load due to irrigation pumps was over 50% of the total load—in some as high as 70%.[15] Perhaps just as dramatically, with the introduction of electric well pumps, washing machines, brood heaters, and other appliances, electrification transformed the non-market but nevertheless productive nature of rural "women's work."[16]

Second- and Third-Tier Cooperatives

In most instances, the new distribution cooperatives depended on unsympathetic IOUs for their wholesale power supply. In an effort to protect their membership, distribution cooperatives federated with one another to create their own capacity through second-tier generation and transmission cooperatives. Over the years, the lure of economies of scale, reliability considerations, and the desire for independence from the price-setting behavior of the for-profit sector in the industry has led G&T

13 "Inexpensive electricity delivered by rural cooperatives presented no competitive threat to investor-owned utilities since they had shown they were unwilling to extend service to rural America, including the High Plains" (Rhodes and Wheeler, 1996, p.312). Nevertheless, as electric distribution cooperatives formed, the IOUs expanded selective efforts to distribute power to rural areas, sometimes in ways that undermined the cooperatives' efforts: "The companies were by no means providing area coverage. Instead they waited to learn where farmers were signing up people for a cooperative, then sent their crews into those areas to build lines in the most thickly settled sections, neatly cutting the heart out of the proposed cooperatives " (Ellis, 1966, p.45).

14 Rhodes and Wheeler, op cit.

15 Rhodes & Wheeler, op cit.

16 Jellison (1993).

systems to consolidate and create *third tier* systems—federations of G&T cooperatives that are themselves federations of distribution cooperatives. Through these arrangements, distribution coops contract indirectly with 3rd level systems for their power, though of course at this remove the transaction that begins to resemble simple trade in commodities, and to lose its distinctive cooperative character.

Second- and third-tier cooperatives from their inception have tended to focus first on the management of large contracts for government hydro-power (through for example the Tennessee Valley Authority, the Western Area Power Administration or other federal power management administrations), and later—as federal projects became fully subscribed and loads continued to grow—on cooperative generation ventures. Increasingly, over time, the cooperatives have signed on as junior partners with IOUs in the development of new generation capacity, typically from coal.

G&Ts typically sell power to the distribution cooperatives under long-term (25-35 years, commonly) exclusive or *all-requirements* contracts that expressly limit the RECs ability to purchase from other sources or to generate its own power. Historically, all-requirements contracts between second- and third-tier G&Ts and their member cooperatives were required by the Rural Electrification Administration, serving as a form of collateral against the REA's below-market financing. From both the benefit and the cost side, the terms of this arrangement are now less favorable for distribution cooperatives: not only are other sources of financing available (particularly in areas with competitive wholesale markets), but the financial risks associated with new environmental regulation—which will legally be passed through onto all-requirements-contracted members—are greater now than ever before.

While accepting the terms of this bargain has effectively protected the RECs from the worst case scenario of being left vulnerable to unchecked IOU exercise of monopoly pricing power, it has done so at the cost of turning away from some of the RECs' own important institutional capital. By ceding control of the transmission and generation of the electricity they purchase for their members, these second- and third-tier contracts relinquished the prerogatives of current members of the distribution cooperatives *as potential producers*.

Some have argued that the RECs are simply consumer cooperatives.[17] While that may be the case, it blurs the distinction between retail distribution cooperatives—which is too narrow a definition to encompass the RECs, either in their origins or in their present incarnations—and, more accurately, special service and supply cooperatives. The latter are much closer cousins to production cooperatives, which would be a necessary framework for individual RECs to embrace if they were to become serious developers of windpower.

IV. REC Natural Advantages

When compared with non-utility independent power producers, the RECs seem objectively to have economic, institutional, and geographic advantages. They have already "organized the need." They also have the load, the infrastructure, and the engineering and financing infrastructure that most competing developers must contract for through the open market. And they are uniquely positioned to implement wind power broadly (not necessarily intensively), with systemic benefits for the electricity grid as a whole. On one level, exploiting a free, local, and inexhaustible resource should appeal strongly to the spirit of self-sufficiency and community development that originally informed the cooperative movement. The fact that the "fuel" is free, clean, and inexhaustible also provides a powerful conservative hedge against both regulatory risk and fuel price risk.

Institutional Assets: As a practical matter, cooperatives stand ready with considerable industry and engineering expertise, if not in-house then at least close at hand. Perhaps even more importantly, they have functioning transmission, distribution, and billing networks in place. Whereas an independent power producer must contract with an existing utility for interconnection and long-term power purchase—often on unappealing terms—the RECs sit behind a unified meter, from which perch they can displace wholesale power purchases directly.

Financing: With zero fuel costs and minimal operation and maintenance, wind turbines are classic examples of high fixed-cost, low marginal cost projects.

17 See for example Hansmann (1996), chapter 9. Hansmann puzzles over the prevalence of cooperatives in rural areas, and their relative absence in urban areas. He argues that "transiency" in cities increases the cost of consumer ownership, partly by complicating the capital accounting, partly by distorting the incentives to monitor management, and partly by (through median voter politics) increasing the likelihood of exploitative pricing at the expense of industrial and commercial users. While these are interesting theoretical considerations, the far simpler answer is that the investor owned utilities *got there first* in the case of the urban areas, because it was profitable for them to do so, whereas the rural areas discovered that provision of the service under remotely acceptable terms would have to be self-organized.

Asynchronous costs and benefits underscore the critical role of financing and interest rates, and here also, the RECs have institutional advantages over many other organizations that might contemplate wind power development. Through the Rural Utilities Service (RUS) of the US Department of Agriculture, the successor to the REA, cooperatives have access to favorable loan rates and guarantees. For renewable generation the terms of RUS finance are even more advantageous. The last two federal energy bills have also been laced with grant funds and other incentives that RECs are well positioned to exploit.

Geography: The geographically dispersed populations served by RECs are well aligned with small-scale, distributed power systems. It has long been understood that distributed generation can yield positive externalities to the rest of the electricity grid. Generation close to load places less demand on scarce long-distance transmission lines, which not only makes the system less vulnerable to cascading power failures, but also can reduce congestion at key nodes of the grid. The economic benefits of stability and of reduced congestion can be substantial.

V. Obstacles to REC Involvement in Wind Development

Despite these apparent advantages, RECs have not led the way on wind power adoption. While among the 1000 or so RECs there are naturally some important differences, the official stance of the national umbrella group NRECA could be fairly characterized as extremely cautious.[18] While there are clearly some implementation obstacles that haven't yet been discussed, NRECA's wariness goes beyond cautioning members not to leap before they look, and approaches something very like a policy of opposition to wind-friendly legislative and regulatory developments.[19] Evidently, NRECA perceives wind power more as a threat than as an opportunity. The question is: Why?

Two distinct issues arise here. First is the question of why individual RECs haven't been more aggressive about developing wind themselves, for their own use. Second is

---

18 NRECA (2003). NRECA (2001).

19 See for example NRECA and APPA (2005), wherein a request is made for rehearing before the Federal Energy Regulatory Commission on FERC's final ruling (Order 661, June 16, 2005) on terms of interconnection of wind energy: "The Final Rule has departed from [the] carefully crafted and balanced approach of the NOPR without reasoned basis or adequate record support. The final result unduly discriminates and creates preferences in favor of wind developers, owners and operators, shifting the costs of maintainng system reliability onto customers and competing generation sources" (p. 2).

the question of why RECs have apparently thrown their lobbying muscle behind a go-slow approach to wind development generally.

As discussed above, existing contractual obligations do mitigate against wind developments within a distribution cooperative. RECs typically have entered into long-term commitments with their power supplier (frequently second-tier G&T cooperatives) to meet all of the cooperative's power needs and have agreed not to develop their own generating capacity.

Federal tax and energy policies also work against the cooperatives in some direct ways. The single-most important federal incentive for wind development in the US is a production tax credit (PTC). This feature of the tax code, along with an accelerated depreciation allowance, privileges for-profit, corporate structures over other ownership models. Distribution cooperatives simply lack the tax appetite to capture these lucrative subsidies.

Nor is regulatory policy ideal for the adoption of wind on the part of RECs. In some regions there remains no incentive whatsoever to build near load. Other regions of the country have adopted congestion pricing, which has begun to create an incentive structure which rewards generation closer to the load. With congestion fees determined in real time, and extremely sensitive to changes in the spatial pattern of generation and load, however, it is difficult to project the savings stream from erecting local generation.[20] And network savings to the grid as a whole—for example from enhanced stability due to reduced need for transmission—will not be captured by the REC that builds local generation.

These three obstacles are not completely insurmountable. Long-term all-requirements contracts do come up for renewal, though generally only on a 25-35 year basis. The disadvantage conferred by federal tax code is an artifact of public policy, as opposed to a fundamental matter of economics. The lack of sufficient incentive in the electricity market for stability-enhancing distributed generation, is likewise an artifact of policy, and therefore subject to change.

20 Nor is there a sufficient historical track record of these prices. The Midwest Independent System Operator (MISO), for example, has run a wholesale day-ahead and real-time market with nodal pricing only since April, 2005. It would be extremely risky to generalize from this limited sample in order to parameterize a model of the cost savings associated with a new generator.

But the fact remains that NRECA, the umbrella organization for RECs, perceives a real threat from wind. This stems from some legitimate concerns about stranded costs, as well as uncertainty about how the restructuring of the electricity industry will ultimately play out.

Restructuring and Stranded Costs: From the 1930s until the late 1970s the electricity industry in the US developed largely on the basis of regulated, vertically integrated monopolies, each with an exclusive geographic footprint. As we have seen, the RECs, in fact, emerged to fill the interstices between the areas which the investor-owned utilities (IOUs) chose to develop and serve. Exclusivity was by design in the case of the IOUs, as the economies of scale in power generation, transmission, and distribution, were thought to render competition economically inefficient. In the case of the RECs, exclusivity was initially by default—due to lack of interest on behalf of for-profit IOUs—but later came to be a kind of shield, protecting against IOU encroachment, or the running of "spite lines" to the more profitable customers in an REC's service area.

The emergence of relatively efficient small-scale generating technologies that used clean, renewable, non-imported fuels, even in the midst of spiking global energy prices, led to a significant regime change in the late 1970s: passage of the Public Utilities Regulatory Policy Act (PURPA) in 1978, included a key provision obliging utilities, for the first time, to receive and pay for electric power produced by "qualifying facilities" (QFs).[21] The Energy Policy Act of 1992 formally set the stage for widespread competition in the generation of electricity, not limited to QFs, and has spawned a flurry of activity to restructure the industry. In some cases the IOUs have been forced to sell off generating capacity to independents, and remake themselves into regulated transmission and/or distribution firms. While the majority of power (about 80% in the Midwest Independent Systems Operator region) is still traded through long-term bilateral contracts, competitive spot- and day-ahead wholesale markets have become commonplace.[22]

---

21 QFs are primarily non-utility generators using renewable energy technologies. PURPA instructs utilities to pay "avoided costs" to the QFs for all power purchases. While the principle is simple (all costs which the QFs' electricity production enabled the utility to avoid—from energy costs to ancillary services costs to reduced congestion costs—should be fully remitted by the utilities to the QFs), in practice it has been the subject of enormous regulatory effort to define.

22 The prices which emerge from these markets are expected to play an increasingly important role as references prices for the bilateral markets.

Subjecting to competition only one stage of production in a vertically integrated monopoly has proven complex. A great deal of regulatory effort has been devoted to the conditions and terms of open access because competition in generation is meaningless without access for the independent power producers to the existing transmission and distribution networks. But while open access makes competitive generation markets possible, it inevitably threatens to strand certain investments, made under the old regulatory regime. Those may take the form of investments in generators which have been rendered uncompetitive by new technologies (but that still must be paid for), or investment in transmission lines, over which the owners now must accept the regulators' terms of access and compensation, or even investment in distribution infrastructure to serve an existing customer, which is suddenly stranded by the customer's decision to generate power behind the meter.

In this brave new restructured world there is uncertainty not only over which assets will be rendered worthless by a new entrant, but also over which entity will be saddled with the costs of providing services that benefit the entire grid. Load balancing, transmission maintenance and upgrade, conservation and demand management, among other ancillary activities, are services essential to the well-being of the system, whose cost could, at least potentially, be thrust upon some hapless distribution cooperative without a means to recover them. "Restructuring" is a high-stakes game. The existing RECs, with their long-term contracts for power and their fixed investment in infrastructure, fear that they may not be positioned to win a game that depends upon agility.

The principal threat to the RECs from the restructuring process in general, and from wind power in particular, involves the obligation to interconnect QFs under PURPA.[23] In what is generally regarded as a positive development, terms (technical and financial) for interconnection are increasingly standardized, through extensive rulemaking dockets at the level of state public utility commissions.

Interconnection entails direct costs, of course, but it also threatens to strand some of the investments of the REC. In the case of a customer interconnecting on-site power, this might strand (in proportion to the customer's power use) the REC's

[23] The difficulties one small QF had in providing wind power in a REC service area is illustrative of the controversy. See Szczepanski (2006).

investment in a second tier G&T cooperative, or other long-term power contract. For the most part, small RECs can apply for a waiver from FERC rules 888 and 889 (which require public utilities to publish open-access, non-discriminatory transmission tariffs), but in cases where the REC does provide some transmission service it might find itself under FERC jurisdiction, obliged to accommodate even wholesale generators who demand interconnection.[24] RECs are correct to fret over stranded costs which wind development, done by others, might impose upon them; but logically they should not allow those misgivings to prevent them from taking advantage of the opportunity to develop the wind in-house.

REC Rate Structures and Wind Power: Most RECs have very simple rate structures: most commonly, REC distribution tariffs include a small fixed charge—often just enough to cover meter reading and billing—and a constant energy charge per kilowatt hour (kWh) consumed. Average fixed costs are recovered by setting the rate per kWh slightly above the average cost of service for customers in the particular rate class. But while such a tariff structure excels at simplicity and ease of implementation, it doesn't serve at all well when the consumer demand is highly variable. Almost all behind-the-meter customer applications of wind will still require some level of service from the utility, and such an arrangement will require the utility to bear real costs that guaranty system adequacy to meet the customer's maximum load at system peak. Such costs cannot be recovered by the standard tariff, since the quantity of kWhs demanded will presumably be *reduced* by the installation of the turbine. Other members of the cooperative would simply have to absorb a larger portion of the fixed costs. The nightmare scenario for RECs is that a vicious circle be put into motion, whereby each customer that opts out imposes costs on remaining members, and thus increases the incentive for them to opt out as well.

Tariff problems are not insurmountable but they are complex. Tariffs which combine peak demand charges with marginal energy charges are more difficult to explain to customers, but nevertheless find wide implementation among IOUs and are not unknown to RECs. While the argument could certainly be made that the costs of a QF opting out should be "socialized" to a level broader than the individual front-line REC, as long as this is not the case, the RECs are legitimately focusing

---

24 See the BVEA case study: NRECA (2003; p.48-51).

attention on ensuring that rates for distributed generation fully protect remaining cooperative members.[25] Ultimately, subtly designed retail tariffs can hedge some of the risk that REC customer demand behavior poses, but the real protection of the load-serving entities requires an architecture of the market as a whole with side payments that adequately compensate the positive externalities of distributed generation.

RECs are potentially in an ideal position to lead on wind power adoption: they have the geography, the expertise, and the "organized need." But they face a dilemma: policies that make wind adoption easier and more economical also make it more likely that the RECs will find themselves forced to buy more power from more QFs, and without some reform of electricity tariffs and market architecture as a whole, such a development would threaten to strand REC investments and shift costs to remaining REC members. Ironically, this casts the RECs in a defensive posture with respect to wind development and distributed generation, and makes of them allies of the IOUs rather than progressive agents of change.

VI. Community Wind

Most of the existing and marginal wind generation capacity in the US can be fairly characterized as "corporate wind," or large-scale projects developed by utility companies, making full use of the accounting and tax advantages, as well as of some limited technological increasing returns to scale. Another niche might be described as "community wind." This has been variously defined, but generally refers to smaller (maximum 10-20 MW) projects—still involving utility-scale turbines—where local ownership is an integral feature. Development of wind on this scale has taken place under a wide variety of business models.[26] In many ways, community wind embodies the spirit of the cooperative movement, including a commitment to local ownership and the potential for democratic (rather than share-based) control. It has been widely argued that community wind is capable of fueling adoption of wind turbine technologies where corporate wind might fail, precisely because by virtue of its investment the local community is less apt to protest any disamenities associated

---

25 NRECA has developed elaborate guidelines for how RECs should attempt to price distributed generation arising from independent producers (see NRECA (2001)). Increasingly, however, the DG tariffs are a matter of PUC policy, rather than individual REC jurisdiction. A recent Minnesota PUC docket (E-999/CI-01-1023, September 2004), for example, standardized the technical and financial terms for non-utility distributed generation connection to the grid, as well as the standard-offer tariffs (purportedly reflecting avoided costs) at which each utility will purchase the power.

26 Bolinger *et al.,* (2004).

with the towers and turbines themselves. Research on community wind has established its greater local economic impact, relative to corporate wind.[27] While community wind has played a large—even dominant—role in European wind development, its role historically in the US has been far more subdued. Not only is community wind small as a percentage of total wind capacity (approximately 2.2%), but it is characterized by stark regional differences, arising in turn from key differences in state regulatory and incentives structures.

The odd mixture of state and federal policies goes a long way towards explaining the location and terms under which community wind has thrived in the US. At the federal level, the key incentives consist of an inflation-adjusted production tax credit of $18/MWh, as well as a 5-year accelerated depreciation allowance. Each of these is clearly more conveniently exploited by corporate investors with tax liabilities sufficient to generate an "appetite" for these shelters. Intermittent, competitively-awarded grant programs and interest rate subsidies at the federal level have also influenced the present value calculations for wind development projects.

Community wind has thrived quite remarkably in the state of Minnesota, while the state of Iowa comes in a distant second. In the former, over 155 MW of installed capacity is already on-line, vending power to utility companies, and at least 460 MW should be developed by 2010.[28] The existence and moderate success of community wind is directly attributable to state policies that accommodate and even privilege the relevant forms of development and ownership. The Minnesota developments have been spurred on by quantitative mandates, imposed by the PUC on the state's largest utility as part of a *quid pro quo* for settlement on an unrelated matter.[29] Additional state policies relevant specifically to community wind include an intermittently available cash production incentive of $.015/kWh directly from the state coffers, payable to subscribed projects under 2 MW, and various capital grant programs.

27 US General Accounting Office (2004). Kildegaard and Myers-Kuykindall (2006).

28 Windustry website, last viewed 9/04/07: *http://www.windustry.org/maps/map_default.htm*

29 Xcel Energy agreed to support development of 425 MW of wind power by 2002, an additional 400 MW by 2006, and an additional 300 MW by 2010, in exchange for permission to store spent nuclear fuel rods in above-ground casks at the Prairie Island nuclear plant. At least 160 MW of this is required to be from projects of 2 MW or less. According to further terms of this arrangement, Xcel must meet Minnesota's Renewable Energy Objective (REO), which ratchets required production from renewables from 1% of retail sales in 2005 to 10% by 2015. Xcel has operationalized its support via a standard 20-year power purchase agreement offering $.033/kWh.

Thus community wind in Minnesota is intermediate between the remarkable level of development in parts of Europe,[30] and the dearth of significant progress in other US states. This measure of success is due to the fact that hard quantitative mandates, standard offer power purchase agreements, and favorable production incentives (federal and state) have combined to make the pricing incentives in the Minnesota system behave quite like the European model until the year 2000, effectively driven by feed-in tariffs. The fact that the glass is also half-empty owes something to the federal incentives that privilege corporate development, as well as to the absence of Danish-style local ownership requirements.

VII. Conclusion

While community wind—a sort of cousin to cooperative wind—has found positive traction and carved out an identity in the industry, the existing rural electric cooperatives have behaved much like the load-serving IOUs: they have been reluctant to purchase wind, wary and unenthusiastic about legislation encouraging wind, and cautious about developing wind farms on their own. Their caution with respect to purchasing wind and to encouraging independent generation is understandable: the brave new world of deregulated electricity markets poses a real threat of stranding existing REC investments, and the load-serving entities are least agile of the players in this particular game—certainly when compared to the high-powered strategic and lobbying muscle of certain energy trading companies. To endure and to thrive in a world where they are obliged to connect wind power, the RECs will at a minimum need to develop rate structures that adequately account for the risks and externalities of QFs and behind-the-meter installations.

Concern over the lack of cooperative wind development is not simply misplaced nostalgia or romanticism. The economic benefits of local ownership are increasingly well documented. There are also excellent reasons to believe that the social costs of contracting for various wind development services will be lower for RECs than for private community wind developers. Measurement of system integration costs is controversial, but there are good reasons to think that a distributed generation model will involve lower social costs than the central-station-power and high-voltage-transmission model toward which the corporate wind developers have gravitated.

30 According to Bolinger (2001) over 80% of Danish and German wind capacity should be classified as community wind.

Despite apparent geographic, institutional, and economic advantages, the RECs' own progress in wind development has been slow. Distribution cooperatives may be limited contractually for the time being, but should be prepared to embrace wind as power purchase agreements permit. As things stand, instead of developing the wind, they are having it developed for them, in some cases out from under them, both by the corporate utilities and by independent QFs. And since current estimates suggest that an upper bound on electricity from wind (for purposes of system-wide stability of the grid) is probably in the 10-15% range, there is a sense in which each megawatt of additional wind developed privately is one that may never be developed cooperatively.

If it is too much to ask that tax law, deregulation policy, and legislative appropriations all take a more considerate look at the plight of the RECs, the best solution may simply be an alliance with community wind developers—be they public or private. There are already plenty of examples of wind development undertaken by municipalities, school districts, economic development authorities, tribes, and other public or quasi-public entities. If it comes to this form of local wind development, the more perfect the match between the jurisdiction that develops the wind and the REC itself, the more closely this arrangement would conform to direct development on the part of the REC, and the greater the prospect that some significant portion of the resource would manifest itself in widely shared benefits at the community level.

(Bart Finzel and Arne Kildegaard are Associate Professors of Economics and Management, University of Minnesota, Morris. The authors can be reached at finzelbd@morris.umn.edu and kildegac@morris.umn.edu respectively).

References

Barsness, Nancy (1979): "The CU Project and its impact on local Cooperatives and Rural Electric Members," in Doyle, Jack, Lines Across the Land, ed. by Vic Reinemer, Washington, D.C.:Environmental Policy Institute, pp 441-472.

Bolinger, Mark (2001): "Community Wind Ownership Schemes in Europe and their Relevance to the United States". LBNL-48357. (Berkeley, CA: Lawrence Berkeley National Laboratory). (*http://eetd.lbl.gov/ea/EMS/reports/48357.pdf*)

Bolinger, M., Wiser, R., Wind, T., Juhl, D., and Grace, R. (2004). *A Comparative Analysis of Community Wind Power Development Options in Oregon.* Portland, Oregon. Manuscript prepared for the Energy Trust of Oregon.

Ellis, Clyde T (1966): *A Giant Step*, New York: Random House.

Folsom, Joe(2003): "Measuring the Economic Impact of Cooperatives in Minnesota," RBS Research Report 200, US Department of Agriculture.

Gardner, Bruce L. (2002): *American Agriculture: How it Flourished and What it Cost.* Harvard University Press.

Hansmann, Henry (1996): *The Ownership of Enterprise.* Harvard University Press.

Jellison,Katherine (1993): *Entitled to Power: Farm Women and Technology,* 1913-1963. Chapel Hill, NC: University of North Carolina Press.

Kercher, Leonard C., Vant W. Kebker, and Wilfred C. Leland, Jr (1941): *Consumer Cooperatives in the North Central States,* ed. by Roland S. Vaile, Minneapolis: University of Minnesota.

Kildegaard, Arne, and Josephine Myers-Kuykindall (2006): "Community vs. Corporate Wind: Does it Matter Who Develops the Wind?" Report prepared for the Institute for Renewable Energy and the Environment, University of Minnesota, and the West Central Regional Sustainable Development Partnership. *http://cda.mrs.umn.edu/~kildegac/CV/Papers/IREE.pdf*

Leiken, Steven (1999): "The Citizen Producer: The Rise and fall of Working-Class Cooperatives in the United States," Chpt. 3 in Furlough, Ellen and Carl Strikwerda, eds., *Consumers against Capitalism?,* New York: Rowman & Littlefield Publishers, Inc.

National Rural Electric Cooperative Association (2001): "Developing Rates for Distributed Generation." (NRECA: Arlington, VA).

National Rural Electric Cooperative Association (2003): "NRECA White Paper on Wind Power." (NRECA: Arlington, VA).

National Rural Electric Cooperative Association, and American Public Power Association (2005): "Request for Rehearing of the National Rural Electric Cooperative Association and the American Public Power Association on the Commission's Final Rule on Interconnection for Wind Energy," Federal Energy Regulatory Commission Docket No. RM05-4-001, July.

O'Brien, Doug (2005): "Legal and Policy Considerations of Investor-Friendly Cooperatives," The National Agricultural Law Center, University of Arkansas. (*http://www.NationalAgLawCenter.org*)

Rhodes, Steven L., and Samuel E. Wheeler (1996): "Rural Electrification and Irrigation in the US High Plains," *Journal of Rural Studies*, V.12, No.3, pp.311-317.

Spear, Roger (2000): "The Cooperativeerative Advantage," *Annals of Public and*

Cooperative Economics, V.71, No.4, (December).

Szczepanski, Carolyn (2006): "A populist movement is afoot in rural Iowa. But will reluctant electric cooperatives and utilities defend their turf and defeat local control?," *Cityview, Central Iowa's Alternative*, February 2.

US General Accounting Office (2004): "Wind Power's Contribution to Electric Power Generation and Impact on Farms and Rural Communities." Report to the Ranking Democratic Member, Committee on Agriculture, Nutrition and Forestry, US Senate. GAO-04-746 (September).

Voorhis, Jerry (1961): *American Cooperatives*. New York: Harper and Brothers.

4

Wind Energy: Issues to Consider

Brian J Frosch and Joe L Outlaw

This policy research report is from the Texas Agricultural and Food Policy Centre, with final results of a research project undertaken by the faculty from The Texas A&M University System. This article discusses Texas leads in America and more than Spain, Germany leads the world in wind power capacity. The article looks at Small Wind Applications for farm and residential use of up to 100 kW generation capacities to Large Wind Applications where landowners in areas with class 3 or higher winds measure output in megawatt where each turbine totes up 3 MW. Additionally Wind Lease Considerations, Wind Lease Glossary, Government Incentives, Availability, Site Selection, Equipment Providers, Utility interconnection, Net Metering and Economic models are touched upon.

Average residential electric prices in Texas have increased approximately 58 percent, or from 8.1 cents per kilowatt hour in 2002 to 12.7 cents per kilowatt hour in 2006. Steadily growing demand for electricity and rising input costs for electric generation are largely responsible for this increase. As producers look to increase

Source: http://www.afpc.tamu.edu © Agricultural and Food Policy Center, College Station, Texas USA. Reprinted with permission.

capacity ahead of demand, increased regulations and environmental concerns over greenhouse gases have opened the door to renewable sources of electricity production, such as wind. According to the World Wind Energy Association (WWEA), 14,900 megawatts (MW) of wind generation capacity were added worldwide in 2006, with the US leading the way adding 2,454 MW.[1] Overall, Germany leads the world in wind capacity while the US has recently surpassed Spain to reach the second position.[2]

In the US, Texas claims the top spot in installed wind capacity with 4,356 MW, or 26 percent of total US capacity, as of the end of the fourth quarter of 2007.[3] Texas accounts for another 1,238 MW of the 3,626 total capacity that is under construction in the US[4]. In addition, Texas ranks second in wind energy potential, meaning a large portion of the state has winds that are suitable for wind energy generation for small farm, ranch, or residential applications or large wind farm applications. Furthermore, federal and state incentives for renewable energy generation have created a favorable environment for investment in wind energy systems. For those landowners who find themselves in a suitable location for wind energy production, the following is a general overview of issues to consider for both small and large wind applications. More detailed information can be found via the links provided.

Small Wind Applications

Landowners looking to add a small wind turbine for farm or residential use are generally referred to as small wind applications, or up to 100 kilowatts (kW) of generation capacity. A typical farm or residential application consists of a 10 kW turbine mounted on an 80 foot tower and costs around $3,000-$5,000 per kW of generation capacity.[5] If a battery storage system is used to store excess electric generation, this generally will double the cost of the system. Federal and state incentives may help to offset the capital cost so it is important to become familiar with those early in the process.

Before embarking on such a project, it is important for landowners to research and review all applicable local, state, and federal laws and regulations. Small wind

1 World Wind Energy Association (WWEA), available online: *http://www.wwindea.org.*

2 West Texas Wind Energy Consortium, "West Texas Wind Wire". Volume 3, Issue 6, November 2007.

3 American Wind Energy Association (AWEA), available online: *http://www.awea.org.*

4 Ibid.

5 Ibid.

systems can be installed as stand alone systems, or as grid connected systems. Grid connected systems generally require the approval of the utility provider. Utilities are required by federal law to purchase production from independent providers at the utility's avoided cost of generation, however, this does not apply to electric cooperatives.[6] Several factors must be examined when considering an investment in a small wind system: wind availability or class, zoning restrictions, generation capacity needed, site selection, and utility interconnection.

First and foremost, does the potential site have sufficient wind availability for a turbine to operate? Generally, small, grid-connected wind applications require a minimum annual average wind speed of 5 meters per second (mps) or close to 11 miles per hour (mph).[7] It should be pointed out that each year technological improvements in the turbines allow for economic efficiency at lower wind speeds. It is best to check your manufacturers' recommendations.

Annual average wind speed data for the US can be found on the National Renewable Energy Laboratory's website. Texas data can be found at the Texas State Energy Conservation Office (SECO) website while more specific weather station locations can be found on the National Climatic Data Center site (websites are listed at the end). For small wind applications, this data is a reliable source to use when determining if your location has sufficient wind to power a turbine. A class 2 wind is generally recommended as the minimum for small wind applications.[8] For a more accurate record of wind data at a specific location, a tower mounted anemometer is the best method to do so. An anemometer is an instrument that is used to collect wind speed and directional data over time, typically a year or more. These systems are generally not cost effective for small wind applications as they generally cost from $14,000 for a 50 foot tower to $25,000 for an 80 foot tower.[9] To help offset this cost, SECO has implemented an anemometer loan program, administered by the Alternative Energy Institute at West Texas A&M. This program provides the data logger, sensor(s), and wiring for the landowner provided tower.[10]

---

6 Ibid.

7 American Wind Energy Association (AWEA), available online: *http://www.awea.org.*

8 Bergey Windpower Company, available online: *http://www.bergey.com/.*

9 AAT Inc., available online: *http://www.aat-solutions.com/?id=19*

10 Texas State Energy Conservation Office, available online: *http://www.seco.cpa.state.tx.us/re_wind_smallwind.htm#loan*

Once it is determined that there is sufficient wind for a turbine, the next step is to review local zoning restrictions, if applicable. Suburban or rural zoning restrictions must be carefully studied with regard to the use of wind turbines, the tower height, and the noise produced. First, one must check zoning restrictions to make certain turbines are allowed or take the necessary steps to get permits. Tower height may be restricted due to nearby airports or to ensure the tower would fall completely on your property in the event of a structural failure. The noise level emitted from the turbine must be in compliance with stated noise restrictions.

Next, an estimate of annual electric needs must be computed to determine optimal generation capacity. This can be accomplished by looking at historical electric bills to determine monthly and/or annual usage. According to Bergey, residential homes typically consume 1,000 to 2,000 kilowatt-hours of electricity per month which translates to a turbine rated between 5-15 kW.[11] They list their 10 kW unit (BWC Excel-S) as the best selling residential unit in the US[12] For those looking at adding a turbine for a specific farm use such as an irrigation pump, turbine capacity would be estimated based on the equipment manufacturer's electric requirements and turbine manufacturer recommendations. Once annual consumption is estimated, investors can research different turbine manufacturers and work with them on selecting a turbine.

When selecting a site, it is also a good time to contact your utility provider to begin determining and preparing for their requirements to connect to the grid. A favorable site relies on a combination of features, including zoning restrictions, topography, and proximity to transmission lines. It is important that the turbine be mounted at least 30 feet higher than existing structures, such as trees and buildings, within 300 feet to ensure consistent wind flow without turbulence. If the potential wind turbine site is near an airport or landing strip, the Federal Aviation Administration has regulations that must be taken into consideration when selecting the tower height. If a connection to the grid is desired, proximity to existing transmission lines is an important consideration that will affect the total investment. Generally, utility companies will extend transmission lines a certain distance at no charge, then a fixed amount per foot thereafter. Selecting a site that is both

---

11 Bergey Windpower Company, available online: *http://www.bergey.com/*.

12 Ibid.

geographically favorable and is in close proximity to transmission lines is ideal. If this is not possible, both factors must be weighed in determining which is most economical or practical for your situation.

Once the site is selected and the turbine is in place, its time to get it online. If the unit is a stand alone system, or will only be connected to a house or specific farm or ranch equipment, the turbine manufacturer should aid in this transition with knowledge and expertise as to appropriate equipment that will be needed, such as an inverter. Those connecting to the grid will need to contact their utility company to aid in the connection. For this type of system, the Public Utility Commission of Texas has a publication, the Distributed Generation Interconnection Manual, which covers safety and technical requirements for installing a distributed generation system.[13] In general, utilities will require some type of interconnection agreement. These agreements generally outline the terms, whereas excess production that goes into the grid will be purchased by the utility provider at a standard price. For those systems with net metering, an agreement will be required as well with the utility provider. Net metering is a simplified method of metering electricity produced and consumed by a home or farm with its own electric generation capacity. Specifically, the meter on this type of system spins backward when the turbine is producing more electricity than the home or farm is using. The consumer is only billed for the net energy consumed during the billing period, essentially receiving the retail value for electricity produced.

Large Wind Applications

Landowners who are in an area with class 3 or higher winds may be seeing an increasing number of large wind developments in their area. Just as for the small wind applications, the wind speeds required for efficient operation are declining. These projects are typically measured in megawatts for the combined capacity of all turbines, where each turbine may have a rated capacity of up to 3 MW. Developers negotiate with landowners for the right to place one or typically multiple turbines on their land. For landowners considering leasing their land for a wind project, there are a number of factors that need to be considered. Although it is not a requirement, it is a good idea for landowners to consult with a lawyer when negotiating or reviewing lease terms and conditions. Some similarities may exist between traditional mineral

13 Public Utility Commission of Texas, available online: *http://www.puc.state.tx.us/electric/business/dg/dg.cfm*

leases and wind leases, however, there are a number of fundamental differences the landowner must be aware of when entering into a wind lease agreement, including, but not limited to, lease terms, royalties, surface rights and other agreements such as easements and meteorological towers.

When preparing to negotiate or examining a contract to lease your property for wind energy, it is important to keep in mind both parties are working toward a mutually beneficial agreement. Consult with neighbors, lawyers, or provided references to get a concept of the general lease terms and conditions in your area. In addition, review any mineral or other leases on the property under consideration for conflicts. It is important that surface rights are upheld with regard to current revenue sources such as ranching or farming. Some types of hunting may be limited or prohibited in the terms of the agreement, such as hunting with a centerfire rifle, as some agreements may limit the amount of deer hunting while others may prohibit it. Those relying on hunting as a primary revenue source may negotiate compensation for lost revenue if the developer does not allow rifle hunting. In general, most wind energy leases have a 20 year term. Royalties from electricity produced typically begin at 2-4 percent and may escalate up to 6 percent over the 20 year term, which translates to around $2,000 and up per turbine per year.[14] Recent (2008) experience in Texas indicates that as the turbines generation capacity increases, the minimum level per turbine increases to around $4,500 per year.[15]

Other agreements considered in addition to the land agreement are surface rights, easements for primary outgoing transmission lines, and/or substations, and a lease for meteorological tower(s). With wind leases, surface rights are typically specified in the agreement, as the lessee has no automatic right to use the surface.[16] If the land allows, transmission lines from each turbine will typically be located underground, along with wiring for turbine controls, to the primary outflow lines. The primary outgoing transmission lines are located above ground to the substation and subsequent main transmission line on the grid. In addition, developers may seek permission to lease land for a meteorological tower to collect wind speed and directional data for the wind development. Either or

---

14 Energy and Environmental Resource Center, "Harvesting the Wind". University of North Dakota, April 2003.

15 Fuqua, James. Personal Communication. May 12, 2008.

16 Fambrough, Judon. "Wind Rights and Wrongs." Tierra Grande. *Journal of the Real Estate Center at Texas A&M University*, April 2008.

both of these may be included in the primary lease agreement or treated as a separate agreement.

Remember, this is not intended to be a comprehensive list of wind lease terms and conditions. Proper research and consultation with professionals experienced in wind lease agreements is a good complement to the negotiation process. An excellent resource for wind lease considerations can be found through the link on the following page. A number of additional resources for landowners researching wind energy are also listed.

(Brian J Frosch is a Research Associate at Texas A&M University (TAMU) and can be reached at bfrosch@tamu.edu

Joe Outlaw is a Professor and Extension Economist in the Department of Agricultural Economics at Texas A&M University. He also serves as the Co-Director of the Agricultural and Food Policy Center (AFPC) at Texas A&M University. The author can be reached at joutlaw@tamu.edu).

Wind Lease Considerations

http://recenter.tamu.edu/pdf/1856.pdf/-

Wind Energy Glossary

http://www.undeerc.org/wind/literature/Wind_ Glossary.PDF

Federal & State Incentives

http://www.dsireusa.org/

http://www.dsireusa.org/library/includes/genericfederal.cfm?Current PageID=1&state =us&ee=1&re=1

http://www.seco.cpa.state.tx.us/re_wind-U.S.incentives.htm

Small Wind

h*ttp://www.seco.cpa.state.tx.us/re_wind_ smallwind.htm*

http://www.eere.energy.gov/windandhydro/windpoweringamerica/pdfs/small_wind/ small_wind_guide.pdf

http://www.eere.energy.gov/windandhydro/windpoweringamerica/pdfs/small_wind/ small_wind_corn.pdf http://www.eere.energy.gov/consumer/your_home/electricity/index.cfm/mytopic=10880

http://www.awea.org/smallwind/

Wind Availability

http://www.seco.cpa.state.tx.us/re_wind_maps.htm

http://www.nrel.gov/wind/resource_assessment.html

http://rredc.nrel.gov/wind/pubs/atlas/maps.html

http://www.nrel.gov/gis/wind.html

http://www.ncdc.noaa.gov/oa/ncdc.html

Site Selection

http://www1.eere.energy.gov/windandhydro/federalwindsiting/

http://www.awea.org/smallwind/toolbox2/INSTALL/evaluate.html

http://www.eere.energy.gov/windandhydro/windpoweringamerica/siting.asp

Equipment Providers

http://www.awea.org/smallwind/smsyslst.html

Utility Interconnection

http://www.awea.org/smallwind/toolbox2/grid_connecting.html

http://www.dsireusa.org/library/includes/incentive2.cfm?Incentive_Code=TX10R&state=TX&CurrentPageID=1&RE=1&EE=1

http://www.puc.state.tx.us/electric/business/dg/dg.cfm

Net Metering

http://www.awea.org/pubs/factsheets/netmetfin_fs.PDF

http://www.dsireusa.org/library/includes/incentive2.cfm?Incentive_ Code=TX02R&state=TX&CurrentPageID=1&RE=1&EE=1

Economic Model

http://www.eere.energy.gov/windandhydro/windpoweringamerica/docs/small_wind_economic_model.xls

References

1. World Wind Energy Association (WWEA), available online: *http://www.wwindea.org.*
2. West Texas Wind Energy Consortium, "West Texas Wind Wire". Volume 3, Issue 6, November 2007.
3. American Wind Energy Association (AWEA), available online: *http://www.awea.org.*
4. Ibid.
5. Ibid.
6. Ibid.
7. American Wind Energy Association (AWEA), available online: *http://www.awea.org.*
8. Bergey Windpower Company, available online: *http://www.bergey.com/.*
9. AAT Inc., available online: *http://www.aat-solutions.com/?id=19*
10. Texas State Energy Conservation Office, available online: *http://www.seco.cpa.state.tx.us/re_wind_smallwind.htm#loan*
11. Bergey Windpower Company, available online: *http://www.bergey.com/.*
12. Ibid.
13. Public Utility Commission of Texas, available online: *http://www.puc.state.tx.us/electric/business/dg/dg.cfm*
14. Energy and Environmental Resource Center, "Harvesting the Wind". University of North Dakota, April 2003.
15. Fuqua, James. Personal Communication. May 12, 2008.
16. Fambrough, Judon. "Wind Rights and Wrongs." Tierra Grande. *Journal of the Real Estate Center at Texas A&M University*, April 2008.

5

Wind Power Today

This paper from the US Department of Energy discusses the plan to provide 20 per cent of the country's electricity supply in the US for which wind capacity has to increase significantly. Incorporating this in the electricity portfolio could avoid significant emission which is equivalent to the amount of carbon produced by the entire transportation sector up to four years. Economic issues faced by rural communities across nations will be addressed and local financial community needs to explore wind development options, benefits, and barriers. Some of the other issues discussed are technology acceptance, rural economic development, and wind power for native populations, wind for schools, environmental assessment, integration of grid wind R&D, and systems integration. Reducing costs through improved technologies and international collaborations are seen as the answer and projections have been made for the next decade.

Leading the Nation's Wind Energy R&D

Wind energy technology has come a long way over the past decade. In 1996, the average utility-scale wind turbine was almost as tall as a 12-storey building and it produced enough electricity to power about 125 average American homes. At the time, these turbines were considered by some to be quite large, but by today's standards, they would be considered small for utility-scale production. The average turbine installed in 2006 (rated at 1.5 MW) was twice as tall as the '96 model. It is almost as tall as the Statue of Liberty and has a rotor large enough to sweep a football field. A 1.5-megawatt (MW) turbine produces enough electricity to power almost 500 homes, and again, that might be considered small when compared to the 3- to 5-MW machines being developed today that will generate enough power for more than 1,300 homes. A 3.6-MW machine has a rotor diameter large enough to park 24 cars in end to end, and a 5-MW machine is as tall as the Space Needle in Seattle, Washington.

The US Department of Energy (DOE) has worked with industry for more than 25 years to bring the technology to where it is today, developing larger machines that are more efficient and that capture more energy from the wind. As the machines have increased in size and performance, the cost of producing energy has dropped—from $0.80 (current dollars) per kilowatt-hour (kWh) in 1980 to about $0.04/kWh today—so that in some areas of the Nation, utility-scale wind power is the most cost-effective form of new generation available.

DOE has also been working to improve the performance and reduce the costs for small and distributed wind energy systems. These systems show great potential for engaging local populations in addressing America's energy future. Advances in small wind technology have produced quieter and more reliable systems that are easier to install and cost less to operate.

The wind energy industry has become the fastest growing utility scale energy resource in the Nation, growing from 1,800-MW of capacity in 1996 to more than 11,600 MW in 2006. 2006 was a record-breaking year with new installations of more than 2,400 MW and a 27% annual growth rate. The new generating capacity installed in 2006 represents a capital investment of almost $4 billion, more than 10,000 new job-years nationwide (10,000 one-year jobs or 1,000 ten-year jobs), and more than $5 to $9 million in annual payments to landowners. The land payments and jobs provide a much needed economic boost to America's struggling rural economies.

As a clean, domestically produced renewable energy resource, wind energy also contributes to our Nation's energy security and environmental quality. The current capacity will generate more than 30 million megawatt-hours (MWh) and displace approximately 18 million metric tons of carbon dioxide per year.

Although 11,600 MW is enough capacity to power about 3 million average homes, it still constitutes a very small share of the total US generation. According to the Energy Information Administration, as of October 2006, wind accounted for only 0.7% of the national electric supply. Coal-fired plants generate the majority (48.5%) of the Nation's electrical energy, followed by natural gas (20.5%) and nuclear plants (19.2%). Conventional hydroelectric, petroleum-fired plants, and other renewables constitute the remaining generation sources.

To accelerate the development and use of advanced clean energy technologies, President George W. Bush launched an Advanced Energy Initiative in 2006. According to the Initiative, areas with good wind resources have the potential to supply up to 20% of the electricity consumption of the United States. The DOE Wind Energy Program is collaborating with industry and stakeholders to analyze credible scenarios for high levels of wind energy use, and determine what actions will best address the technology, market, and policy challenges to maximizing the Nation's opportunity for harnessing its immense wind resources. To provide 20% of the Nation's electricity supply, US wind capacity would have to increase from its current 11,600 MW to more than 325,000 MW. Incorporating this amount of wind generated electricity in the Nation's electricity portfolio could avoid emission of 3,500 million metric tons of carbon equivalent through 2050—equivalent to the amount of carbon produced by the entire transportation sector over 3-1/2 years. This would also lead to approximately $332 billion in economic investment and more than 3,725,000 full-time equivalency job years for construction and plant operation, largely focused in rural areas.

To provide greater support for the President's initiative, in 2006, the DOE Wind Energy Program shifted the emphasis of its activities to accelerate the market penetration of wind technology. Shifts in program activities include:

1. Supporting the development of 20% US energy by wind.
2. Increasing the program's efforts to overcome near-term deployment barriers to wind projects.

3. Collaborating with the DOE Office of Electricity Delivery and Energy Reliability to ensure that wind energy is appropriately represented in grid expansion and modernization efforts around the Nation and conducting outreach to Federal, state, and local organizations and utilities.
4. Expanding work in the area of turbine performance and reliability to mitigate risk to investors, developers, and operators.
5. Expanding program activities in the distributed wind technology market sector (residential, farm, small business).

Wind Energy Program Support

The DOE Wind Energy Program is one of 10 clean energy technology programs within the Department of Energy. It is managed by program staff at DOE Headquarters in Washington, D.C., and its Project Management Center (PMC) in Golden, Colorado, manages the financial assistance, provides program support, and conducts outreach activities. The program provides funding to a wide range of public and private sector partners, with primary focus on the National Renewable Energy Laboratory's (NREL) National Wind Technology Center (NWTC) near Boulder, Colorado, and Sandia National Laboratories (SNL) in Albuquerque, New Mexico. NREL and SNL conduct wind energy research with industry partners and researchers from universities nationwide to advance wind energy technologies. Each laboratory is extensively equipped with a unique set of skills and capabilities to meet industry needs.

As the lead research facility for the program, NREL's NWTC conducts research across the complete spectrum of engineering disciplines that are applicable to wind energy, including: atmospheric fluid mechanics and aerodynamics; dynamics, structures, and fatigue; power systems and electronics; and wind turbine engineering applications. The center also provides design reviews and analysis; dynamometer, field, and blade testing services; and field verification for wind turbines that range in size from 400 watts to 2.5 MW. The NWTC is the Nation's only wind energy technology test facility accredited to International Electrotechnical Commission (IEC) standards. Industry partners may use the center's facilities to conduct atmospheric, static-strength, and fatigue tests on turbines and components, including its 2.5-MW dynamometer, to conduct lifetime endurance tests on a wide range of wind turbine drivetrains and gearboxes. In addition, the NWTC completed construction of a

225-kW dynamometer in 2005 that will aid development of advanced generators and power electronics for small wind systems. The NWTC also has two permanently installed advanced wind turbines to test new control schemes and equipment, and sites for testing industry prototype wind turbines.

SNL specializes in all aspects of wind-turbine blade design and system reliability. Activities at SNL focus on reducing the cost of wind generated electricity and improving the reliability of systems operating nationwide. Research disciplines include: materials, airfoils, stress analysis, fatigue analysis, structural analysis, and manufacturing processes. By partnering with both universities and industry, SNL has advanced the state of knowledge in the areas of materials, structurally efficient airfoil designs, active-flow aerodynamic control, and sensors. Researchers at the laboratory are currently investigating integrated blade designs where airfoil choice, blade planform, materials, manufacturing process, and embedded controls are all considered in a system perspective. By collaborating with operators, developers, and manufacturers, SNL evaluates known reliability problems and develops tools and methods to anticipate and investigate future reliability issues.

Technology Acceptance

In support of Advanced Energy Initiative objective to expand the use of wind energy, the Wind Energy Program is increasing its efforts to overcome near-term deployment barriers to wind by enhancing public acceptance, promoting supportive public policies, engaging key stakeholders, and addressing siting and environmental issues.

In 1999, only four states boasted more than 100 MW of installed wind capacity. By the end of 2006, 16 states had more than 100 MW and six more states are expected to reach that capacity by the end of 2007. The goal of the DOE Wind Powering America (WPA) project is for 30 states to have 100 MW of wind installed by 2010.

To achieve its goal, WPA supports the formation of state wind working groups, providing stakeholders with timely information on the current state of wind technology, economics, state wind resources, economic development impacts, and policy options/issues. Group members include landowners and agricultural sector representatives, utilities and regulators, colleges and universities, advocacy groups, and state and local officials. In 2006, WPA launched four new state wind working groups in Illinois, Indiana, Missouri, and New Jersey, bringing the total number of state wind working groups to 29. WPA also supported events in 11 states and convened its 5th annual All-States Summit in Pittsburgh, Pennsylvania. The summit provided participants with an opportunity to share strategies and lessons learned and to visit with experts on topics such as avian and wildlife issues, siting, transmission, community wind, small wind, Native American projects, operating impacts, utility myths, regulators, radar, interconnection, and wind resources and mapping.

Rural Economic Development

Rising fuel costs, low commodity prices, and a lack of jobs are just a few of the economic issues faced by rural communities nationwide. To address these issues, WPA works with rural community leaders, US Department of Agriculture local and national representatives, state and local officials, the Farm Bureau, the Farmers' Union, representatives of growers associations, agricultural schools, and the local financial community to explore wind development options, benefits, and barriers. Achieving the goals of WPA during the next 20 years will create $60 billion in capital investment in rural America, provide $1.2 billion in new income for farmers and rural landowners, and create 80,000 new jobs.

Wind Power for Native Americans

The United States is home to more than 700 Native American tribes located on 96 million acres (39 million hectares), much of which have excellent wind resources that could be commercially developed to provide electricity and revenue to the reservations. Before these resources can be fully realized, many issues need to be resolved. These include the lack of wind resource data, tribal utility policies, sovereignty, perceived developer risk, limited loads, investment capital, technical expertise, and transmission to markets. To support the development of Native American wind resources, WPA provides a wide range of technical assistance and outreach activities to more than 20 tribes from 13 states. To help tribes understand their wind resource and potential development options, WPA administers a Native American Anemometer Loan Program. In 2006, WPA helped install four 40- to50-m (130 to 165 ft) towers and one 20-m (65 ft) tower. Installation of two to four more of the tall towers is anticipated in 2007.

WPA also provides wind energy training for Native Americans through the DOE-supported Wind Energy Applications and Training Symposium (WEATS) at the NWTC. Training sessions in the 2006 symposium included Wind Applications, Wind Fundamentals, Small Wind (on- and off-grid applications), Site Selection and Wind Resource Assessment, Land Agreements/Environmental Review, Permitting, Interconnection and Transmission, and Wind Integration. Participants also toured the NWTC and the Ponnequin Wind Farm.

Wind for Schools

At a grassroots level, WPA is engaging rural America in a discussion of wind energy while developing a knowledge base through its Wind for Schools (WfS) project. The objectives of the project are to:

1. Engage rural school teachers and students in wind energy education
2. Equip college students in wind energy applications and education to provide the growing US wind industry with interested and equipped professionals
3. Introduce wind energy on a small scale in rural communities, starting a discussion of the benefits and issues in using wind energy.

To accomplish its objectives, the WPA team at NREL assists schools with the installation of a small wind turbine through a coordinated community effort. Team members include a WfS facilitator within each state; a wind application center at a state-based university or college to provide technical assistance; a school, science teacher, school administration, and community to host or own the wind turbine; a green tag marketer to assist with the sale of the green attributes of the turbine to defray system costs; a wind turbine manufacturer to provide the wind turbine system; the local utility or energy cooperative; and the state energy office.

WPA launched it first WfS project in Colorado in 2006 and plans to replicate the process in Nebraska, Kansas, South Dakota, Montana, and Idaho in 2007-2008.

Environmental Assessment

WPA also works with universities and non-government organizations to address wind turbine siting issues such as aesthetics, radar interference, and wind-wildlife interactions. In 2006, the program worked with AWEA, the National Wind

Coordinating Committee (NWCC), and other federal agencies on wind power-radar interaction issues that affected more than 1000 MW of planned installations.

Wind power-radar interaction issues gained national attention in 2006 due to the potential for radar operations to be affected by wind turbines. Interference occurs when radar signals are reflected back by wind turbines causing clutter on the radar screens. In July, more than 100 experts, including representatives from AWEA, DOE, the Department of Defense, and the Federal Aviation Administration attended a Wind Power and Radar Issue Forum brief convened by the NWCC to discuss the influence of wind energy on aviation radar and possible mitigation strategies. This collaboration and follow-on interaction helped facilitate the approval of 950 MW of wind projects.

To help resolve wind-wildlife interactions, the program supported two collaborative efforts, the Grassland Shrub Steppe Species Collaborative—a 4-year effort to study the impact of wind turbines on prairie chicken habitats in Kansas—and the Bat and Wind Energy Collaborative that investigates bat and wind turbine interaction. In addition, the NWCC hosted its sixth Wildlife Research Meeting in Texas. The purpose of the meeting was to bring participants up-to-date on research being conducted to understand the interaction of birds, bats, and other wildlife with wind energy development, examine what has been learned about ways to minimize or mitigate wind energy's impacts on wildlife, and identify gaps in knowledge and research needs.

Facilitating Integration of Wind on the Grid

Systems Integration

The natural variability of the wind resource can present challenges to grid system operators and planners with regard to managing regulation, load following, scheduling, line voltage, and reserves. While the current level of wind penetration in the United States and around the world has provided substantial experience for successful grid operations with wind power, many grid operators need to gain a better understanding of the impacts of wind on the utility grid before they can feel comfortable increasing the percentage of wind in their energy portfolios. The goal of the program's systems integration research is to address interconnection impacts, electric power market rules, operating strategies, and system planning needed for wind energy to compete without disadvantage to serve the Nation's energy needs.

In 2006, the Wind Energy Program conducted a number of studies to provide states and utilities with the information and tools they need for wind energy development. For example, the program provided extensive GISbased wind resource and transmission data to help the Western Area Governors Association (WGA) and the Clean and Diversified Energy Advisory Committee (CDEAC) to identify 30 GW of clean power that could be developed by 2015. In 2006, the CDEAC Wind Task Force developed a set of supply curves based on this data. The findings of the study indicate that the wind resource in the WGA region is more than enough to economically achieve the WGA 30-GW target for clean energy development.

Additional 2006 system integration activities included a study that provided the Western Farmers Electric Cooperative with a systems integration and wind power data analysis, a wind integration study for the state of Minnesota, and a study for Xcel Energy in Colorado that provided the company with the data it needs to assess the technical and economic impact of adding a significant amount of wind generation to its energy portfolio.

NREL Supplies WAPA with Wind Farm Training

Simulator

The Western Area Power Administration (WAPA) provides electric power system operations training to power system operators and dispatchers throughout the United States and Canada at its Electric Power Training Center (EPTC) in Golden, Colorado. A significant portion of this training is performed with the Miniature Power System (MPS), an actual power system consisting of three synchronous generators (scaled up to 500 MW total), five loads, two ties to the western electrical grid, and a simulation of more than 500 miles of transmission lines. To enable WAPA to integrate wind energy into its training program, NREL's NWTC supplied WAPA with a wind farm simulator in 2006 that successfully simulated power delivery to the MPS grid from a time-series file of real wind-farm data for a 50-MW wind farm.

Enhancing Critical Energy Infrastructure

Transmission is a key energy infrastructure element critical to tapping our national wind resource and moving electricity to market, much as the interstate highway system does for the Nation's transportation needs. Much of the Nation's best wind resources cannot be tapped to meet our increasing energy demands without new transmission system capacity. The development of new transmission is challenged by many regulatory, jurisdictional siting, and cost allocation barriers. The program is working with state and Federal energy offices as well as regional organizations and utilities to support appropriate representation of wind energy characteristics and opportunities in energy infrastructure planning processes underway across the nation.

The development of new transmission corridors requires the coordination of many different organizations and groups from the Federal, regional, state, and local levels. Upgrading the Nation's transmission system, like upgrading the interstate highway system, will have substantial costs and will cross many organizational boundaries.

The Wind Energy Program is working closely with the DOE Office of Electricity Delivery and Energy Reliability to effectively coordinate the Department of Energy's contributions to the transmission planning efforts. This joint program effort will focus on linking remote regions with low-cost wind power to urban load centers, allowing thousands of homes and businesses access to abundant renewable energy.

Reducing Costs through Improved Performance

Large Wind Technologies

Although research efforts for the past two decades have led to dramatic reductions in the cost of wind energy, continued incremental improvements to wind turbine performance will lower system costs even further while improving system integration and enhancing technology acceptance.

The Wind Energy Program focuses its cost energy reduction efforts on improving wind turbine components. For example, gearboxes comprise 35% to 40% of the total wind turbine system cost. To help industry identify opportunities for improved gearbox design, the program initiated a long-term industry collaboration. NREL sponsored a drivetrain workshop in 2006 to jointly identify research needs under a multilateral industry-driven test program at the NWTC. (For more information, contact Sandy Butterfield at 303-384-6902 or *sandy_butterfield@nrel.gov).*

Extending the fatigue life of system components like the drivetrain, blades, and tower will play an important role in reducing system costs. As wind turbines become larger and taller, they become more flexible and susceptible to fatigue. To design fatigue-resistant wind turbines, the program is investigating ways to gain better control of the way the components interact and move. Control systems that regulate turbine power and maintain stable closed-loop behavior in the presence of turbulent wind inflow are critical to today's large wind turbine designs. NREL is developing and testing control systems that maximize energy capture while reducing structural dynamic loads that cause turbine parts to wear out thus increasing the cost of operation and maintenance.

To better understand improvement opportunities for wind turbine availability, SNL hosted a Wind Turbine Reliability Workshop in Albuquerque, New Mexico in October, 2006. More than 90 participants, representing wind farms, service companies, consultants, manufacturers, universities, and laboratories listened to presentations on topics ranging from hardware reliability (gearboxes, generators, pitch systems, blades, and condition monitoring) to stakeholder perspectives (owners, operations, maintenance, and user groups). The workshop also addressed planning for the characterization and reduction of operating and maintenance costs, data information sharing, and the establishment of user groups to address pressing common issues. SNL is planning another technical session in September 2007, and is currently working with the American Wind Energy Association (AWEA) Operations and Maintenance Working Group and others to begin a systematic data collection effort to collect, analyze, and report on issues affecting turbine reliability and availability. (For more information contact Roger Hill, (505) 844-6111, *rrhill@sandia.gov).*

To help increase the performance of wind turbine blades, program researchers have tested new blade designs and materials at NREL's NWTC for the past decade using fatigue and static strength tests. However, the rapid growth in wind turbine size has recently outstripped the capacity of the blade test facilities. In 2006, the program announced a CRADA seeking partners to design, construct, and assist in operating wind turbine blade test facilities capable of testing blades up to at least 70 m (230 ft) in length. The Wind Energy Program will contribute capital equipment and provide NREL staff and expertise to help develop and operate the facility. Massachusetts and Texas were chosen as the two finalists from six applications. The partners in Massachusetts include the Massachusetts Technology Collaborative, the University of Massachusetts, and the Massachusetts Executive Office of Economic Development. In Texas, the Lone Star Wind Alliance, led by the University of Houston and the Texas General Land Office, is partnering with Texas A&M University, Texas Tech University, University of Texas-Austin, West Texas A&M University, Montana

State University, Stanford University, New Mexico State University, Old Dominion University, and the Houston Advanced Research Center.

Distributed Wind Technologies

Distributed wind systems have been traditionally defined as wind turbines rated at 100 kW or less installed at remote locations. The Wind Energy Program has supported efforts to increase the reliability and performance of distributed wind turbines with a goal of producing electricity at between 10 and 15 cents/kWh in a Class 3 wind resource (5.3 m/s at 10 m) by 2007. As achievement of the program's distributed wind technology (DWT) goal draws near, the program is expanding its activities and its definition for distributed applications to include wind turbines that are installed remotely or connected to the grid at the distribution system, including behind the customer meters. An independent assessment of the various segments of the distributed wind market is in progress.

The program is also continuing its efforts to increase the performance and reliability of small wind turbines. In 2007, the program will launch an effort to establish a small turbine testing and certification program. The program will partner with industry to test a number of small turbines to International Electrotechnical Commission (IEC) and draft AWEA standards. The project will provide high-quality, detailed, and independent test results and allow small businesses the opportunity to earn a certification granted by an independent certification body. The certification body is in the process of being formed by the Interstate Renewable Energy Council (IREC)—the Small Wind Certification Corporation.

Successful 2006 Small Wind Projects

Southwest Windpower conducted performance optimization and blade-fatigue tests at the NWTC on its new Skystream wind turbine. The 1.8-kW turbine, developed in partnership with DOE, won the Best of What's New Award from *Popular Science Magazine* and was listed as a best invention for 2006 by *Time* magazine. The new turbine has fully integrated electrical components, costs less, is easier to install, and more quiet to operate.

Northern Power Systems (NPS) is reconfiguring its 100-kW cold weather turbine for agricultural and community applications in temperate climates. The company began building its new machine in 2007 and plans to start testing the prototype at the NWTC before the end of the year. The machine will cost less to produce, and it shows good potential for filling a market gap in mid-sized wind turbines.

Windward Engineering produced a new 4.25-kW machine called the Endurance. The turbine is sized to offset the energy consumption of an average US home (~11,000 kWh/yr) when installed in a Class 3 wind regime (5 m/s at a height of 10m). It employs an induction generator to simplify grid compatibility and a brake that is capable of stopping the rotor on command in any wind condition—a unique feature for a small wind system. Windward used off-the-shelf components from other industries to reduce system cost. The Endurance is currently being tested at the NWTC to IEC standards for duration, power performance, and acoustics.

Wind Energy for the Next Decade

For the past two years, the wind industry has enjoyed record-breaking growth, and industry experts predict that with the extension of the PTC through 2008, the next two years will be record-breakers as well. The challenge for industry is to maintain the long-term wind energy growth required to fulfill the expectations of the President's Advanced Energy Initiative. AWEA and the Wind Energy Program have formed a collaborative with 75 participating companies and organizations to evaluate credible scenarios for providing 20% of US electric demand with wind and identify strategic actions that include:

- Participating in collaborative partnerships with industry that will result in turbine technology with higher reliability, expanded performance and a more competitive cost of energy
- Increasing its outreach and education efforts
- Investigating ways to increase and make transmission more accessible
- Addressing environmental issues
- Providing support for national and state policies that enable robust growth of wind energy.

The program also supports the development of expanded testing capabilities to support larger wind turbine R&D. The CRADA issued in 2006 that seeks partners to build a blade test facility capable of testing blades up to at least 70 m (230 ft) in length is one area targeted for expansion. The program is also exploring options for construction of a new drivetrain test facility of at least 11 MW in capacity.

For distributed wind technologies, the program will support potential markets by investigating applications such as off-grid water pumping for crop irrigation, residential-scale wind turbines, community wind, and hybrid wind/diesel applications.

To ensure long-term wind growth, the program is also investigating emerging applications for wind energy such as offshore installations, hydrogen production, and the production and delivery of clean water.

Hydrogen production offers an opportunity for wind to provide low-cost, clean energy for the transportation sector. NREL, in partnership with Xcel Energy, launched a wind-to-hydrogen demonstration project at the NWTC that will investigate the potential of using wind energy to produce hydrogen. The project will use two wind turbines (a 10-kW and a 100-kW), two proton-exchange membrane electrolyzers, and one alkaline electrolyzer to produce hydrogen from water. The hydrogen will be compressed and stored for later use in a hydrogen internal combustion engine where it will be converted into electricity and fed into the utility grid during peak demand hours. (For more information visit *http://www.nrel.gov/hydrogen/proj_wind_hydrogen.html).*

The program is also investigating wind energy applications that can ease the demands on the Nation's water resources. As the US population grows, it places greater and greater demands on water supplies, wastewater services, and the electricity needed to power the growing water services infrastructure. Water is also a critical resource for thermoelectric power plants. Wind offers an energy source that uses limited water when compared to thermoelectric generation, and it can play a role in supplying energy for municipal water supplies and processes.

All of these applications present new challenges to the wind community, and cost and infrastructure barriers are expected to be significant. The program's vision is that this evolution pathway will begin to have an impact on the marketplace in the post-2020 timeframe.

Wind Energy Web Sites

US Department of Energy Wind and Hydropower Technologies Program *http://www1.eere.energy.gov/windandhydro/*

American Wind Energy Association *http://www.awea.org/windmail@awea.org*

National Renewable Energy Laboratory National Wind Technology Center *http://www.nrel.gov/wind/*

Sandia national Laboratories *http://www.sandia.gov/wind/*

National Wind Coordinating Committee *http://www.nationalwind.org/*

Utility Wind Integration Group *http://www.uwig.org*

Wind Powering America *http://www.eere.energy.gov/windandhydro/windpoweringamerica/*

Successful 2006 LWT Projects

One program R&D project that shows potential for demonstrating a significant increase in overall system performance is the 2.5-MW Liberty wind turbine developed by Clipper Windpower. Clipper completed its prototype in 2005 after only 3 years of R&D. The new machine's innovative distributed-path powertrain design incorporates four permanent-magnet generators, and advanced variable-speed controls. According to US Department of Energy Secretary Samuel W. Bodman, "Clipper's Liberty Turbine is not only one of the most advanced wind turbines ever produced, it may well be the most efficient wind turbine in the world." Successful field-tests conducted by Clipper with assistance from NREL and intensive component testing at the NWTC helped Clipper put the Liberty series turbine into production in the summer of 2006. Clipper's 2006 transaction announcements represent firm commitments of 875 MW of turbines and more than 5,000 MW of contingent orders for delivery through 2011.

Northern Power Systems (NPS) produced an award-winning power electronics package that can be scaled for use in a wide range of wind turbines, from small to multimegawatt systems. According to NPS, the new converter improves wind turbine reliability, energy capture, and grid performance. The project team was chosen by the American Wind Energy Association for its 2006 Technical Achievement Award. Tests completed in 2006 on both the converter and a 1.5-MW direct-drive generator, also developed with program support, demonstrated high-quality power output.

Knight & Carver is developing a 27.5-m (90-ft) replacement blade for a 750-kW turbine. The "STAR" (which stands for sweep twist adaptive rotor) blade is the first of its kind ever built. Its most distinctive characteristic is a gently curved tip, which prompts the blade to respond to high winds such that adverse loads are attenuated. This allows the blade length to be extended with no weight penalty and augments energy capture in low-wind-speed resource areas.

Global Energy Concepts (GEC) worked with program researchers to fabricate a 1.5-MW, single-stage drivetrain with a planetary gearbox and a medium-speed, permanentmagnet generator. The simple gearbox design and moderate-sized generator show potential for reducing tower-head weight and drivetrain costs. The company completed initial testing of this drivetrain at NREL's 2.5-MW dynamometer test facility. The generator is currently being upgraded, and a second phase of testing is planned for 2007.

Genesis Corporation is testing a new tooth form for gearboxes that promises major improvements in power density while reducing the costs of these devices. The company completed the first round of testing with positive results and is now working to refine its design through further targeted testing.

International Collaborations

International Research

The Wind Energy Program supports international wind energy research efforts as a member of the International Energy Agency (IEA) Wind Energy Executive Committee and by providing operating agents for several IEA Tasks. The United States participates in the five tasks listed below, is the operating agent for three of the tasks, and provides technical experts for the Topical Expert meetings held under Task 11: Base Technology Information Exchange.

- Task 19: Wind Energy in Cold Climates
- Task 20: HAWT Aerodynamics and Models from Wind Tunnel Measurements – Operating Agent
- Task 21: Dynamic Models of Wind Farms for Power Systems Studies
- Task 23: Offshore Wind Energy Technology and Deployment – Operating Agent
- Task 24: Integration of Wind and Hydropower Systems – Operating Agent
- Task 25: Power System Operation with Large Amounts of Wind Power

(For more information on IEA activities, visit the IEA web site at *www.ieawind.org*).

International Standards

NREL also plays an active role in the development of international standards by working with AWEA and the International Electrotechnical Commission Working Groups. International standards provide a critical link to cutting edge research which forms the basis for and harmonization of international design requirements.

NREL's participation in the working groups provides consistent representation for US industry, thus ensuring that standards to do not impede international industry development and trade opportunities while ensuring that environmental, safety, and health interests of industry employees, utility personnel, and the general public are maintained. NREL participates in the following active IEC Working Groups:

IEC 61400-1, Wind Turbine Safety and Design Requirements

IEC 61400-2, Small Turbine Safety and Design Requirements

IEC 61400-3, Offshore Design requirements

IEC 61400-4, Wind Turbine Gearbox Requirements

IEC 61400-11, Wind Turbine Noise Measurement

IEC 61400-12, Wind Turbine Power Performance Measurement

IEC 61400-21, Power Quality

IEC 61400-22, Certification Requirements

IEC 61400-23, Blade Structural Testing

IEC 61400-24, Lightning Protection Guidelines

IEC 61400-25, Communications and SCADA

6

Whither Wind?

Charles Komanoff

This article is a philosophical and learned presentation on large wind farm siting from a man whose life's work has been fighting fossil fuels and machines powered by them. The engagement is over the disagreement about whether big windmills belong and whether they belong at all then and that too, with the backdrop of how the world needs to reduce CO_2 emissions by at least 50 percent over the next few decades. The new ethic for wind power siting is discussed in the light of the fact that it should be much easier to comprehend the impact of wind farm development than the less tangible losses from a warming earth. The discussion regarding various issues like 'why me, why my ridgeline, my seascape, my viewshed' are examined with a suggested ranking matrix of locales to cope with the location issues to ensure their very existence.

It was a place I had often visited in memory but feared might no longer exist. Orange slabs of calcified sandstone teetered overhead, while before me, purple buttes and burnt mesas stretched over the desert floor. In the distance I could make out southeast Utah's three snowcapped ranges—the Henrys, the Abajos, and 80 miles to the east, the La Sals, shimmering in the blue horizon.

Source: www.motherearthnews.com © Charles Komanoff. Reprinted with permission.

No cars, no roads, no buildings. Two crows floating on the late-winter thermals. Otherwise, stillness.

Edward Abbey's country. But my country, too. Almost 40 years after Abbey wrote *Desert Solitaire,* 35 since I first came to love this Colorado River plateau, I was back with my two sons, who were 11 and 8. We had spent four sun-filled days clambering across slickrock in Arches National Park and crawling through the slot canyons of the San Rafael Reef. Now, perched on a precipice above Goblin Valley, stoked on endorphins and elated by the beauty before me, I had what might seem a strange, irrelevant thought: I didn't want windmills here.

Not that any windmills are planned for this Connecticut-sized expanse—the winds are too fickle. But wind energy is never far from my mind these days. As Earth's climate begins to warp under the accumulating effluent from fossil fuels, the increasing viability of commercial-scale wind power is one of the few encouraging developments.

Encouraging to me, at least. As it turns out, there is much disagreement over where big windmills belong, and whether they belong at all.

Why Wind Farms?

Fighting fossil fuels, and machines powered by them, has been my life's work. As an energy analyst, I can tell you that the science on global warming is terrifyingly clear: To have even a shot at fending off climate catastrophe, the world must reduce carbon dioxide (CO_2) emissions by at least 50 percent within the next few decades. If poor countries are to have any room to develop, the United States—the biggest emitter by far—needs to cut back by 75 percent.

Although automobiles, with their appetite for petroleum, may seem like the main culprit, the No. 1 climate change agent in the United States is actually electricity. The most recent inventory of US greenhouse gases found that power generation was responsible for a whopping 38 percent of CO2 emissions. Yet the electricity sector may also be the least complicated to make carbon free. Approximately three-fourths of US electricity is generated by burning coal, oil or natural gas. Accordingly, switching that same portion of US electricity generation to nonpolluting sources such as wind turbines, while simultaneously ensuring that our ever-expanding arrays of lights, computers and appliances are increasingly energy efficient, would eliminate 38 percent of the country's CO2 emissions and bring us halfway to the goal of cutting emissions by 75 percent.

To achieve that power switch entirely through wind power would require 400,000 windmills rated at 2.5 megawatts each, by my calculations. To be sure, this is a hypothetical figure, since it ignores such real-world issues as limits on power transmission and the intermittence of wind, but it's a useful benchmark just the same.

What would WIND FARMS entail?

To begin, I want to be clear that the turbines I'm talking about are huge, with blades up to 165 feet long mounted on towers rising several hundred feet. Household wind machines such as the 100-foot-high Bergey 10-kilowatt BWC Excel with 11-foot blades, the mainstay of the residential and small business wind turbine market, may embody democratic self-reliance and other "small is beautiful" virtues, but we can't look to them to make a real dent in the big energy picture.

What dictates the supersizing of windmills are two basic laws of wind physics: A wind turbine's energy potential is proportional to the square of the length of the blades, and to the cube of the speed at which the blades spin. I'll spare you the math, but the difference in blade lengths, the greater wind speeds higher off the ground, and the sophisticated controls available on industrial-scale turbines all add up to a market-clinching 500-fold advantage in electricity output for a giant General Electric or Vestas wind machine.

How much land do these industrial turbines require? The answer turns on what "require" means. An industry guideline is that to maintain adequate exposure to the wind, each big turbine needs space around it of about 60 acres. Since 640 acres make a square mile, those 400,000 turbines would need 37,500 square miles, or roughly all the land in Indiana or Maine.

On the other hand, the land actually occupied by the turbines—their "footprint"—would be far, far smaller. For example, each 3.6-megawatt Cape Wind turbine proposed for Nantucket Sound will rest on a platform roughly 22 feet in diameter, implying a surface area of 380 square feet—the size of a typical one-bedroom apartment in New York City. Scaling that up by 400,000 suggests that just six square miles of land—less than the area of a single big Wyoming strip mine—could house the bases for all of the windmills needed to banish coal, oil and gas from the US electricity sector.

Of course, erecting and maintaining wind turbines also can necessitate clearing land: Ridgeline installations often require a fair amount of deforestation, and then there's the associated clearing for access roads, maintenance facilities, and the like. But there are also now a great many turbines situated on farmland, where the fields around their bases are still actively farmed. Depending, then, on both the particular terrain and how the question is understood, the land area said to be needed for wind power can vary across almost four orders of magnitude.

WIND POWER: A Fractious Debate

Similar divergences of opinion are heard about every aspect of wind power. Big wind farms kill thousands of birds and bats ... or hardly any, in comparison to avian mortality from other tall structures such as skyscrapers. Industrial wind machines are soft as a whisper from a thousand feet away, and even up close their sound level would rate as "quiet" on standard noise charts ... or they can sound like "the shrieking sound of a wild animal," according to one unhappy neighbor of an upstate New York wind farm.

Some of the bad press is warranted. The first giant wind farm, comprising 6,000 small, fast-spinning turbines placed directly in Northern California's principal raptor flyway, Altamont Pass, in the early 1980s rightly inspired the epithet "Cuisinarts for birds." The longer blades on newer turbines rotate more slowly and thus kill far fewer birds, but bat kills are being reported at wind farms in the Appalachian Mountains; as many as 2,000 bats were hacked to death at one 44-turbine installation in West Virginia. And as with any machine, some of the nearly 10,000 industrial-grade windmills now operating in the United States may groan or shriek when something goes wrong.

At the same time, however, there is an apocalyptic quality to much anti-wind advocacy that seems wildly disproportionate to the actual harm, particularly in the overall context of not just other sources of energy but modern industry in general. New York state opponents of wind farms call their Web site "Save Upstate New York." In Massachusetts, a group called Green Berkshires argues that wind turbines "are enormously destructive to the environment," but does not perform the obvious comparison to the destructiveness of fossil fuel-based power. Although the intensely controversial Cape Wind project "poses an imminent threat to navigation and raises many serious maritime safety issues," according to the anti-wind Alliance to Protect Nantucket Sound, the alliance was strangely silent when an oil barge bound for the region's electric power plant spilled 98,000 gallons of its deadly, gluey cargo into Buzzards Bay in 2003.

Of course rhetoric is standard fare in advocacy, particularly the environmental variety with its salvationist mentality—environmentalists always like to feel they are "saving" this valley or that species. You can spend hours sifting through the anti-wind Web sites and find no mention at all of the climate crisis, let alone wind power's potential to help avert it.

In fact, many wind power opponents deny that wind power displaces much, if any, fossil fuel burning. This notion is mistaken. It is true that since wind is variable, individual wind turbines can't be counted on to produce on demand, so the power grid can't necessarily retire fossil fuel generators at the same rate as it takes on windmills. The coal- and oil-fired generators will still need to be there, waiting for a windless day. But when the wind blows, those generators can spin down. That's how the grid works; it allocates electrons.

What about the need to keep a few power stations burning fuel so they can instantaneously ramp up and counterbalance fluctuations in wind energy output? The grid requires this ballast, known as spinning reserve, in any event both because demand is always changing and because power plants of any type are subject to unforeseen breakdowns. The additional variability due to wind generation is slight —wind speeds don't suddenly drop from strong to calm, at least not for every turbine in a wind farm and certainly not for every wind farm on the grid.

With very few exceptions, then, wind output can be counted on to displace fossil fuel burning one for one. No less than other nonpolluting technologies such as bicycles or photovoltaics, wind power is truly an anti-fossil fuel.

What Windmills Signify

I made my first wind farm visits in the fall of 2005, to the 20-windmill Fenner Windpower Project and the seven-windmill Madison Windpower Project, both located in Madison County, N.Y. It was windy, though not unusually so, according to the locals. All 27 turbines were spinning, presumably at their full 1.5-megawatt ratings. For every hour in full use, each windmill was keeping a couple of barrels of oil, or an entire half-ton of coal, in the ground. Of course wind turbines don't generate full power all the time because the wind doesn't blow at a constant speed. The Madison County turbines have an average annual output rate of 28 percent, meaning that over the course of a year, they generate between one-fourth and one-third of the electricity they would produce if they always ran at full capacity. But that still means an average 2,500 hours per year of full output for each turbine. Multiply those hours by the 27 turbines at Fenner and Madison, and a good 160,000 barrels of oil or 40,000 tons of coal were being kept underground by the two wind farms each year.

The windmills, spinning at 15 revolutions per minute—that's one revolution every four seconds—were clean and elegant in a way that no oil derrick or coal dragline could ever be. The nonlinear arrangement of the Fenner turbines situated them comfortably among the traditional farmhouses, paths and roads, while at Madison, a grassy hillside site, the windmills were more prominent but still unaggressive. The windmills didn't seem like a violation of the landscape. The turning vanes called to mind a natural force—the wind—in a way that a cell phone or microwave tower, for example, most certainly does not.

They were also relatively quiet. My sound readings, taken at distances ranging from 100 to 2,000 feet from the tower base, topped out at 64 decibels and went as low as 45—the approximate noise range for a small-town residential cul-de-sac.

Thinking back on that November day, I've come to realize that a windmill, like any large structure, is a signifier. Cell phone towers signify the intrusion of quotidian life—the reminder to stop at the 7-Eleven, the unfinished business at the office. The windmills I saw in upstate New York signified, for me, not just displacement of destructive fossil fuels, but acceptance of the conditions of inhabiting the Earth.

What about the argument that the potential energy produced by wind turbines could be saved instead through energy-efficiency measures? Examples include swapping out incandescent light bulbs in favor of compact fluorescents, replacing inefficient kitchen appliances and extinguishing "vampire" loads by plugging watt-sucking electronic devices into on-off power strips. If this notion sounds familiar, it's because it has been raised in virtually every power plant dispute since the 1970s. But the ground has shifted, now that we have such overwhelming proof that we're standing on the threshold of catastrophic climate change.

Those power plant debates of yore weren't about fuels and certainly not about global warming, but about whether to top off the grid with new supply or energy saved through conservation. The energy arena of old was local and incremental. The new one is global and all-out. With Earth's climate, and the world as we know and love it, now imperiled, topping off the regional grid pales in comparison to the task at hand. In the new, ineluctable struggle to rescue the climate from fossil fuels, efficiency and "renewables" (solar, biomass and wind) must all be pushed to the max.

A New Ethic for WIND POWER Siting

Part of the problem with wind power, I suspect, is that it's hard to weigh the effects of any one wind farm against the greater problem of climate change. It's much easier to comprehend the immediate impact of wind farm development than the less tangible losses from a warming Earth.

Intruding the unmistakable human hand on any landscape for wind power is, of course, a loss in local terms, and no small one. The inevitable access roads for erecting and serving the turbines can be damaging ecologically as well as symbolically. In contrast, you will feel few benefits of the wind farm in a tangible way. If the thousands of tons of coal a year that your wind farm will replace were being mined a mile from your house, it might be a little easier to take.

If Congress enacted an energy policy that harnessed the spectrum of cost-effective energy efficiency together with renewable energy, thereby ensuring that fossil fuel use shrank starting today, a windmill's contribution to climate protection might actually register, providing psychic reparation for an altered viewshed. If carbon fuels were taxed for their damage to the climate, wind power's profit margins would widen. And surrounding communities could extract bigger tax revenues from wind farms, which could go to a new high school, or land acquired for a nature preserve.

It's very human to ask, "Why me? Why my ridgelinee, my seascape, my viewshed?" These questions have been difficult to answer; there has been no framework—local or national—to guide wind farm siting by ranking potential wind power locales for their ecological and community suitability. That's a gap that the Appalachian Mountain Club (AMC) is trying to bridge, using its home state of Massachusetts as a model.

According to AMC research director Kenneth Kimball, who heads the project, Massachusetts has 96 linear miles of "Class 4" ridgelines, where wind speeds average 14 mph or more, the threshold for profitability with current technology. Assuming each mile can support seven to nine large turbines of roughly 2 megawatts each, the state's uplands could theoretically host 1,500 megawatts of wind power. (Coastal areas such as Nantucket Sound weren't included in the survey.)

Kimball's team sorted all 96 miles into four classes of governance—Appalachian Trail corridor or similar lands where development is prohibited; other federal or state conservation lands; Massachusetts open space lands; and private holdings. Then they overlaid these with ratings denoting conflicts with recreational, scenic and ecological values. The resulting matrix suggests the following rankings of wind power suitability:

1. **Unsuitable:** Lands where development is prohibited (Appalachian Trail corridors, for example) or "high conflict" areas: 24 miles (25 percent)
2. **Less than Ideal:** Federal or state conservation lands rated "medium conflict": 21 miles (22 percent)
3. **Conditionally Favorable:** Conservation or open space lands rated "low conflict," or open space or private lands rated "medium conflict": 27 miles (28 percent)
4. **Most Favorable:** Unrestricted private land and "low conflict" areas: 24 miles (25 percent)

Category 4 lands are obvious places to look to for wind farm development. Category 3 lands could also be considered, says the AMC, if wind farms were found to improve regional air quality, were developed under a state plan rather than piecemeal, and were bonded to assure eventual decommissioning. If these conditions were met, then categories 3 and 4, comprising approximately 50 miles of Massachusetts ridgelines, could host 400 wind turbines capable of supplying nearly 4 percent of the state's annual electricity—without grossly endangering wildlife or threatening scenic, recreational or ecological values.

Whether that 4 percent is a little or a lot depends on where you stand and, equally, on where we stand as a society. You could call the 400 turbines mere tokenism against our fuel-besotted way of life, and considering them in isolation, you'd be right. But you could also say this: Go ahead and halve the state's power usage, as could be done even with present-day technology, and "nearly 4 percent" doubles to 7 percent to 8 percent. Add the Cape Wind project and other offshore wind farms that might follow, and wind power's statewide share might reach 20 percent, the level in Denmark.

Moreover, the windier and emptier Great Plains states could reach 100 percent wind power or higher, even with a suitability framework like the AMC's, thereby becoming net exporters of clean energy. But even at 20 percent, Massachusetts would be doing its part to displace a portion of the 75 percent of US electricity generated by fossil fuels. If you spread the turbines needed to achieve that goal across all 50 states, you'd be looking to produce roughly 800 megawatt-hours of wind output per square mile—just about what Massachusetts would be generating in the above scenario.

So goes my notion, anyway. You could call it wind farms as signifiers, with their value transcending energy-share percentages to reach the realm of symbols and images. That is where we who love nature and obsess about the environment have lost the high ground, and where Homo americanus has been acting out his (and her) disastrous desires—opting for the "manly" SUV over the prim Prius, the macho powerboat over the meandering canoe, the stylish halogen lamp over the dorky compact fluorescent.

Throughout his illustrious career, wilderness champion David Brower called upon Americans to "determine that an untrammeled wildness shall remain here to testify that this generation had love for the next." Now that all wild things and all places are threatened by global warming, that task is more complex.

Could a windmill's ability to "derive maximum benefit out of the site-specific gift nature is providing—wind and open space," in the words of aesthetician Yuriko Saito, help Americans bridge the divide between pristine landscapes and sustainable ones? Could windmills help Americans subscribe to the "higher order of beauty" that environmental educator David Orr defines as something that "causes no ugliness somewhere else or at some later time"? Could acceptance of wind farms be our generation's way of avowing our love for the next?

I believe so. Or want to.

— Adapted from a longer article of the same name, which originally appeared in Orion Magazine—a publication that combines creative ideas and practical solutions to reconnect human culture with the natural world.

(For more information on Orion, call (888) 909-6568 or visit *http://www.orionmagazine.org/*).

(Charles Komanoff is Director of KomaÂ-noff Energy AssociÂ-ates. He has written extensively about energy, economics and the environment, as well as pedestrian and bicyclist rights in New York City. The author can be reached at kea@igc.org).

7

Wind Energy: A Promising Future

Deepika M G

The latest technological advancement and research in the field of wind energy paints a very promising picture. A few groups working in the US, the Netherlands and Canada are on their way to set up 'wind farms' which would have installations towering 9 km up in the sky. As per their estimates, 1% of wind energy at high altitudes can meet all the energy demands of the planet. This would come as a breath of fresh air to power-deficit countries such as India, where growth in wind energy for the last few years has been quite promising.

If India has to sustain its attractive growth of 8 to 9% per annum in GDP it is inevitable that it exploits the alternative sources of energy. The power GDP elasticity in India is estimated to be roughly 1.6% signifying the importance of this sector to the overall growth of the economy. India has been for long and continues to be a power deficit country. If the demand for power is growing at the rate of 8 to 9%, the supply has been growing at a rate of 4% causing a wide deficit. Exploitation of the alternative sources of energy, especially solar and wind energy, is foreseen as some solution to this problem. The growth in the production of wind energy for the past few years in India and some parts of the world is quite promising.

The Current Indian and Global Scenarios

Wind is the fastest growing energy source in the world growing at a rate of 30% per annum. India now ranks fourth in the world in terms of wind energy production. The world scenario of wind energy depicts that Germany has the highest installed capacity of 17,000 MW followed by Spain, the US, India, Denmark and other countries summing up to 48,451 MW of total production. Global wind energy production is expected to reach 1,17,412 MW in 2009 with Asia (India and China) having the highest growth next only to Australia. The installed capacity currently in India stands at around 5,340 MW. But this is far from 45,000 MW, which has been estimated by Ministry of New and Renewable Energy in India as India's actual potential. However, the growth in this sector in the last decade has been quite spectacular. In March 1996, the installed capacity was worth of 733 MW whereas in 2006 it has grown to 5,340 MW. At present, the contribution from the renewable energy sources, solar and wind is quite meager at 5% of the total power produced in India. Suzlon Energy Ltd., Enron India Ltd., Shriram EPC Ltd., Southern Wind Farms Ltd. and Vestas Wind Technology India Ltd., are some of the major players owning windmills in India.

Other than the fact that wind is a renewable source of energy there are many boons associated with this form of energy. Fossil fuels which are a source of non-renewable sources of energy are depleting at a faster rate. The future of the energy sector would, therefore depend on the performance of the renewable sources of energy. It is expected that in another 40-50 years fossil fuels are likely to reach their maximum potential. Hazardous waste emission is nil, which is quite exorbitant in producing thermal or hydro sources of energy. As compared to solar energy, the alternative to wind under renewable energy sources, wind energy is cost-efficient with a very low operating cost. The cost of production is roughly five cents per unit of power produced which is 12 cents for solar. Estimates show that both the capital cost and operating cost are likely to asymptotically decline by the next decade due to the likely advancement in technology. Wind turbines are much easier to construct, the gestation period being quite short. Jobs and employment opportunities are high in this sector. The implicit costs such as the damage to environment or the overexploitation of the natural resources are minimal. On the other hand, an implicit advantage is that the land used for wind farms can be simultaneously used for other agricultural purposes too.

Technological Advancements

The extent of power generated through wind largely depends on the speed of the wind and the capacity of the installed wind turbines. Turbines may vary in size with the capacity of 1 KW structure to large machines with the capacity to generate 2 MW. One estimate says that at full capacity a wind turbine operates at a wind speed of 25-30 miles per hour. Even in India, wind energy technologies have tremendously advanced where large capacity wind turbines in the range of 1.25 to 1.65 MW are manufactured. Wind resources in India at about 30 metres above the ground are largely found in the states of Karnataka, Maharashtra, Andhra Pradesh, Tamil Nadu, Madhya Pradesh and some parts of Rajasthan, West Bengal and Orissa. The gross potential is estimated to be the highest in the states of Andhra Pradesh, Gujarat and Karnataka. Tamil Nadu has been the leading state with the installed capacity of 2,893 MW as on March 2006. Wind energy received a boost in Tamil Nadu some time back when the Ministry for Textiles agreed to include wind farms in its Technology Upgrading Fund Scheme, which gave a subsidy of 5% in the rate of interest on capital borrowed for upgrading textile mills, which was an incentive for several textile mills in the state to set up wind energy farms. Power and utility giants such as GE and Siemens are into the manufacturing of wind turbines. Wind turbines being modular in nature have huge export potential since they can be shipped in the CKD mode. Wind energy sector is also highly prone to scale economies. Assuming the same wind speed, a wind mill which is large in size is more economical than a small one. The future is very bright for wind energy largely due to the expected advancement in technology.

The latest advancements in technology and the research being carried out in the field of wind energy paint a very promising picture of harnessing energy through wind which is a renewable source of energy. A few groups working in the US, the Netherlands and Canada are all on the road to setting up wind farms which would have installations towering 9 kms up in the sky in the tropopause what is called the jet stream or corridor of high velocity winds. As per their estimates 1% of wind energy at high altitudes can meet all the energy requirements of the planet. This has been the brainchild of Bryan Roberts, an Engineer in Australia who has worked on how to extract energy using a fleet of whirring gyromills at more than 4 kms above the ground. David Shepard, President of the group working in Coronado, California, says, "If we are able to tap 1% of the wind energy at very high altitudes that would be enough to supply all the world's energy needs. But the whole concept is, however, very challenging. It involves getting the massive windmills ideally 7 to 9 kms up and then connect a cable down for its use."

The Suzlon Way

Suzlon Energy Ltd., Asia's largest wind power company and fifth largest in the world has currently 6% of the global share and 47% share in the Indian market. It has a presence in over eight states in India with over 30 wind farms. A company that was set up as a subsidiary to Suzlon Energy A/s in Denmark has established offices in Chicago, Beijing, Melbourne, Germany and the Netherlands. Suzlon's strategy aims at supply of components of windmills with high quality and strengthening of overseas R&D units. Its technology is advanced and is at par with global standards. Its strategic approach has been to reduce cost and dependence through backward integration. Its acquisition of Hansen Transmission and RE Power led to strong forward and backward linkages in the supply chain. Its promotional strategy includes participation in various said energy trade fairs and use of electronic media advertisements, thus acting as a role model for other players in the sector.

A Bright Future for India

According to the estimates made by the Indian Wind Energy Association, the capacity from wind will exceed 5,000 MW in less than two years up from 3,595 MW. The forecast made by Ministry of Non-Conventional Energy Sources (MNES) says that by 2012 India would have an additional power capacity of 1,00,000 MW of which the share of renewable energy would go up to 10%, of which wind is likely to contribute 75% of the renewable power produced. The huge potential for wind energy can facilitate India to surpass the small European countries which are now the leading producers of wind energy in the world. Germany is now the largest producer, producing wind energy worth 18,000 MW out of the world's 48,000 installed capacity.

The installed capacity for India is roughly contributing to only 10% of world capacity. This is expected to grow to a large extent, outdoing other countries. The demand from India is likely to increase for the simple reason that the major wind producing countries in Europe are expected to stagnate in the immediate future due to lower scope for expansion given their small geographic territory, unless there is a quick breakthrough in the technology. In Europe, the capacity is likely to grow only by 17% in 2009 and in the US by 23%, both being much lower than what is expected in Asia.

According to MNES, India has witnessed a very high growth rate in terms of capacity installations of wind energy. According to the American Wind Energy Association, global electricity consumption is expected to double between 2002 and 2030 with very high growth rates in India and China. Almost all countries of the world are adopting strategies used for promotion of renewable energy sources for power generation. The Government of the Netherlands has set the goal of 10% contribution of renewable energy in total energy supply by the year 2020. In India, many fiscal and financial incentives are created by MNES, which has set a target of 10% share of renewable energy in the power generation. But this is unlike the situation in Europe where it is compulsory for the power sector to generate certain portion of power from renewables to reduce the CO2 levels and also the existence of pollution tax on fossil fuels. MNES has issued guidelines on the cost at which utilities should buy power from renewable power producers. 14 major state utilities have introduced policies for renewable power projects along the guidelines issued by MNES. Tax holiday on profits is extended for 10 years and concessions on customs duty on imports of inputs are provided. It is felt that India could be one of the very attractive hubs for manufacturing and marketing of wind energy in the world, especially in the Asia-Pacific region.

This is predicted for the reason that India is a good storehouse of R&D facilities for product development and improvement. India has the advantage of possessing good trained manpower in the field of wind energy and also an exclusive center for the purpose of training.

(Deepika M G is Faculty Member at the Icfai Business School, Bangalore. The author can be reached at mgdeepika@gmail.com).

Institutional Support for Wind Energy in India

– Aarti Chirag Bhoorat

The Ministry of New and Renewable Energy (MNRE), India, develops renewable forms of energy like wind energy. There are various promotional incentives provided by the Central Government in the form of, Concession on import duty of wind turbine parts, excise duty reliefs, Loans from Indian Renewable Energy Development Agency Limited (IREDA), Income Tax holiday as provided for conventional power projects, etc.

Private players like Suzlon Energy play a vital role in developing wind energy in India. Suzlon Energy is coming up with a windpark in Dhule, Maharashtra which is being considered as the largest in Asia. Its capacity would be over 1,000 MW by the time it is completed.

Various projects that are being carried out in the states of India with good wind potential are below here:

State-wise Wind Power Installed Capacity In India

State	Gross Potential (MW)	Total Capacity (MW) till 31.03.08	Technical Potential (MW)
Andhra Pradesh	8275	122.5	1750
Gujarat	9675	1252.9	1780
Karnataka	6620	1011.4	1120
Kerala	875	10.5	605
Madhya Pradesh	5500	187.7	825
Maharashtra	3650	1755.9	3020
Rajasthan	5400	538.8	895
Tamil Nadu	3050	3873.4	1750
West Bengal	450	1.10	450
Others	2990	3.2	-
Total (All India)	**45195**	**8757.2**	**12875**

Source: Indian Wind Energy Association (InWEA).

Section II

Technology Considerations

A Review of NDT Techniques for Wind Turbines

M A Drewry and G A Georgiou

Wind machines and traditional 'Dutch' windmills preceded electricity supply and were used for grinding grain. They were always attended, sometimes inhabited and, largely, manually controlled. They were integrated within the community, designed for frequent replacement of certain components and efficiency was of little importance. Wind turbines have been in use since 1941. The function of a modern power-generating wind turbine is to generate high quality, network frequency electricity. There is information and advice on the Non-Destructive Testing (NDT) of wind turbines in Europe and internationally, but because mass production of wind turbines is fairly new, no manufacturing standards have been set yet. There is also a need for European standards, in the testing, certification and accrediation of turbines and components. The main objectives in this paper are to review the current state of NDT of wind turbines at manufacture and in-service, and to establish the most promising NDT methods for detecting flaws of most concern.

Source: http://www.atypon-link.com/BINT/doi/abs/10.1784/insi.2007.49.3.137 © The British Institute of Non-Destructive Testing. Reprinted with permission. "This paper was first published in Insight – Non-Destructive Testing and Condition Monitoring (The Journal of The British Institute of Non-Destructive Testing), Volume 49, Number 3, March 2007, pp137-141, and is published here with the kind permission of The British Institute of Non-Destructive Testing and the authors."

1. Introduction

Historically, wind machines were used for grinding grain in Persia as early as 200 BC. Wind turbines have been in use since 1941 when the world's first megawatt-size wind turbine was connected to the local electrical distribution system in Vermont, USA. The function of a modern power-generating wind turbine is to generate high quality, network frequency electricity. To give an example of the scale of the industry today, there are over 35,000 wind turbines worldwide and the UK has 40% of Europe's wind resource.

The main aim of this paper is to provide a review of the current state of NDT of wind turbines, based on published evidence, at the time of manufacture and at the time of periodic inspection.

Whilst it is well known that Acoustic Emission (AE) is successful in detecting and monitoring flaws in wind turbines, it is not strictly an NDT method. Nevertheless it is considered in this review.

2. Manufacturers

Germany, Spain and Denmark accounted for almost 80% of the wind power capacity installed in Europe in 2003[(1)]. Germany continues to be the leader in terms of cumulative and annual megawatt (MW) wind turbines installed. There has also been a huge growth in the Spanish market in recent years but Denmark continues to dominate the manufacturing side of the industry worldwide. Nine of the top ten turbine manufacturers are European and wind energy is an outstanding European success story, with European companies manufacturing more than 90% of the turbines sold worldwide in 2002.

3. Types of Wind Turbines

Wind turbines can be separated into two general types based on the axis about which the turbine rotates. Turbines that rotate around a horizontal axis are most common. Vertical axis turbines are less frequently used. Wind turbines can also be classified by the location in which they are to be used. There are onshore and offshore wind turbines.

3.1 Horizontal Axis

Horizontal Axis Wind Turbines (HAWT) have the main rotor shaft and generator at the top of a tower, and must be pointed into the wind by some means. Small turbines are pointed into the wind by a simple wind vane, while large turbines generally use a wind sensor coupled with a servomotor. Most have a gearbox too, which turns the slow rotation of the blades into a quicker rotation that is more suitable for generating electricity.

3.2 Offshore

The fact that water has less surface roughness than land means the average wind speed is usually higher over open water. This allows offshore turbines to use relatively shorter towers above the water surface, making them less visible. The offshore environment is, however, more expensive compared to onshore for the following reasons:

- Offshore towers are generally taller when the submerged height is also included.
- Offshore foundations are more difficult and more expensive to build.
- Power transmission is through undersea cable, which is more expensive to install.
- The offshore environment is also corrosive and abrasive.
- Repairs and maintenance are much more difficult.
- Offshore wind turbines are outfitted with extensive corrosion protection measures like coatings and cathodic protection.
- The stresses on offshore wind turbine towers are also greater due to tidal stress.

4. Turbine Design and Construction

Most wind turbines have upwind rotors that are actively yawed to preserve alignment with wind direction. The three-bladed rotor is the most popular and, typically, has a separate front bearing with a low speed shaft connected to a gearbox which provides an output speed suitable for a four-pole generator (see Figure 1). Commonly, with the largest wind turbines, the blade pitch will be varied continuously under active control to regulate power at the higher operational wind speeds (furling). Support structures are most commonly tubular steel towers tapering in some way, both in metal wall thickness and in diameter from tower base to tower top. Epoxybased resin systems dominate the market in blade manufacture and carbon fibre reinforcement is increasingly used in big blades.

In 2006, the focus of attention is on technology around and above the 2 MW rating and commercial turbines now exist with heights over 100 m and rotor diameters up to 100 m. Designs with *variable pitch* and *variable speed* dominate the market while *direct drive* generators are becoming more prevalent.

Figure 1: Wind Turbine Open Nacelle
(From Danish Wind Industry Association)

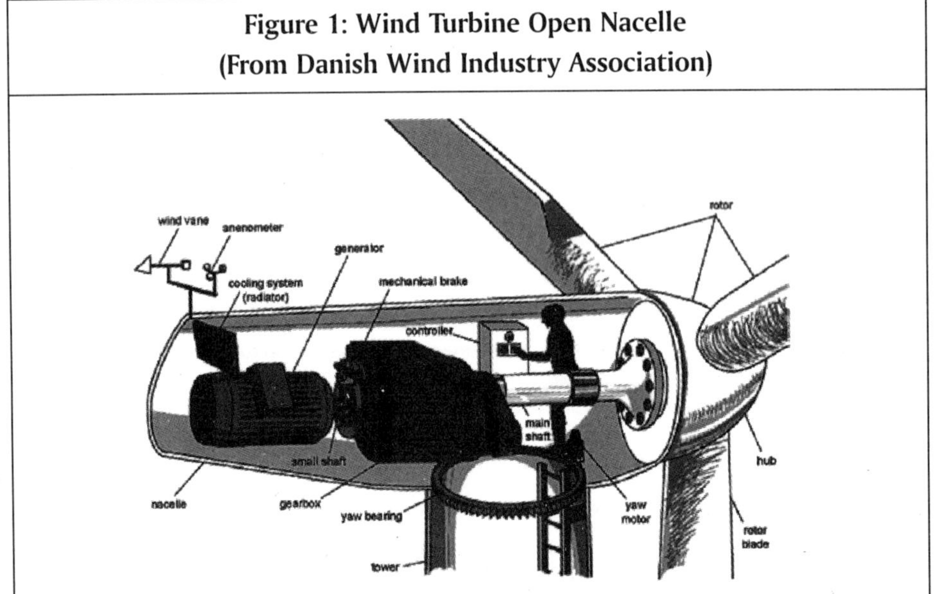

5. Flaws and NDT at Manufacture

Because mass production of wind turbines is fairly new, no manufacturing standards have been set yet. Efforts are now being made in this area on the part of both the government and manufacturers.

While wind turbines on duty are relied on to work 90 percent of the time, many structural flaws are still encountered, particularly with the blades.

Cracks sometimes appear soon after manufacture. Mechanical failure, due to alignment and assembly errors, is common. Electrical sensors frequently fail because of power surges. Non-hydraulic brakes tend to be reliable, but hydraulic braking systems often cause problems.

5.1 Manufacturing Flaws on Turbine Blades

Manufacturing flaws can cause problems during normal operation. For example, blades can develop cracks at the edges, near the hub or at the tips (Figure 2). Fibreglass rotor blades are regarded as the most vulnerable components of a wind turbine(2).

Figure 2: Sketch of Section of Blade (Local Coordinates)

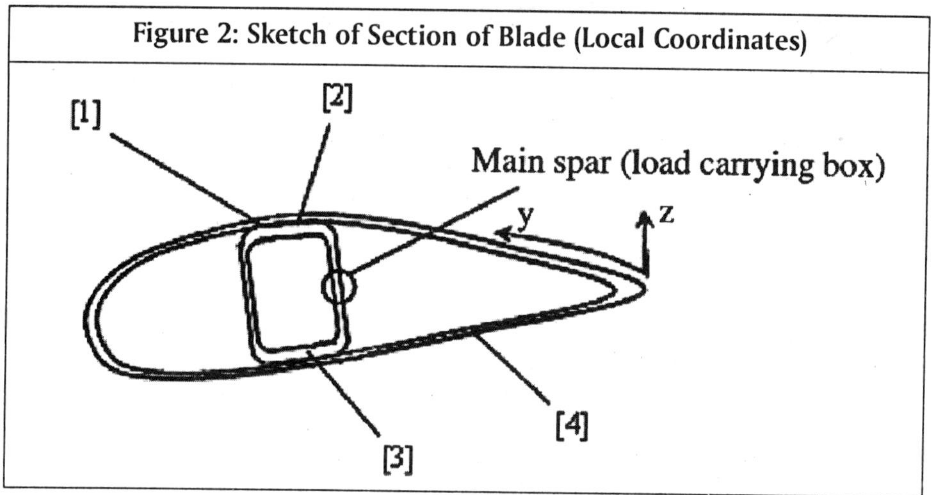

Typical manufacturing flaws on the blades may be summarised as delaminations (Figure 3), adhesive flaws and resin-poor areas. Here are some specific flaws at particular locations:

1. Skin/adhesive: this is bad cohesion between the skin laminate and the epoxy or the epoxy is missing.
2. Adhesive/main spar: this is when there is no cohesion between the adhesive and the main spar.
3. Delamination in main spar laminate.
4. High damping in skin or main spar laminate, which could be caused by porosities or change of thickness of laminate.

Figure 3: Delamination of the Blades

5.2 Manufacturing Flaws on the Tower

A tubular tower is made of a lot of sheets of iron that are welded together. A flange is welded onto each end of the sections. The welding is checked thoroughly using ultrasonic NDT.

5.3 NDT Techniques at Manufacture

5.3.1 Visual Inspection

Relatively advanced NDT testing methods are used to examine rotor blades. The methods employed include penetrant testing and visual inspection with the use of miniature cameras or endoscopes.

At present, it appears that mechanical components aren't tested at manufacture but the cause of their damage can be determined. For example, endoscopes are used for the visual inspection of planetary gear transmissions. However, damaged components are usually examined in a materials laboratory.

5.3.2 Ultrasonic NDT

An ultrasonic test can be carried out to investigate if any damage is present in the wind turbine blade. Ultrasonic inspection reveals these flaws quickly, reliably and effectively and is the most often used non-destructive composite inspection method in industry. The main advantage of ultrasound scanning is that it enables us to see beneath the surface and check the laminate for dry glass fibre and delamination.

5.3.3 Tap Test

The tap test can be used to verify some of the results from the ultrasonic test and it is also a good method to discover irregularities in the structure. The method is based on the fact that the sound emitted when knocking on the structure changes when the thickness or material type changes or when porosities are present. It could also be caused when there is a disbond between the skin laminate and the main spar. There are three types of tap testing equipment; a manual tapping hammer, the 'Woodpecker' portable bondtester and the Computer Aided Tap Tester (CATT) system.

All the automated tap methods have the advantage that they can produce a print of the damaged area, which is both useful and a permanent record of the damage found. All the tapping methods work well for thin laminates, honeycomb structures and other sandwich panels but are not so effective on thicker parts.

5.3.4 Infrared Thermography

The adhesive joints are critical points in the blade structure. That is why they are inspected with particular care. Infrared (IR) scanners are used to examine the blade throughout its length, measuring exactly the same points each time. The scanner is able to see through the laminate and check the adhesive joint. It records temperature differences in the adhesive, possibly identifying flaws, and takes a series of pictures. If there are any doubts, a point can be highlighted and later analysed using electronic image processing. If flaws are found, they can almost always be repaired immediately.

6. Flaws and NDT In-Service

As the force of the wind is so irregular, the driving mechanism of a wind turbine is subject to much greater dynamic loads. Virtually all components of a wind turbine are subject to damage, including everything from the rotor blades to the generator, transformer, nacelle, tower and foundation.

Wind turbines do have regular maintenance schedules in order to minimise failure. They undergo inspection every three months, and every six months a major maintenance check-up is scheduled. This usually involves lubricating the moving parts and checking the oil level in the gearbox. It is also possible for a worker to test the electrical system on site and note any problems with the generator or hook-ups.

6.1 Flaws on Turbine Blades

In-service flaws have been identified in the following report: Risø-R-1391(EN) 'Identification of Damage Types in Wind Turbine Blades Tested to Failure' Christian P. Debel, AFM. ISBN 87-550-3178-1; ISBN 87-550-3180-3 (Internet) ISSN 0106-2840.

6.2 Flaws on the Tower

For wind turbine towers, wind load is regarded as the main load. The analysis of damage in most towers shows that it occurs under a wind of medium intensity, and the reason is fatigue failures (Figure 4). The number of sections which are dangerous from the point of view of fatigue is determined during the design of the structure.

Figure 4: Collapsed Wind Turbine Tower Due to Fatigue Cracks and Wind Loading Combination

6.3 Flaws on Rotor Bearings

Cyclic stresses fatigue the blade, axle and bearing material, and were a major cause of turbine failure for many years. When the turbine turns to face the wind, the rotating blades act like a gyroscope. As it pivots, gyroscopic precession tries to twist the turbine into a forward or backward somersault. For each blade on a wind generator's turbine, precessive force is at a minimum when the blade is horizontal and at a maximum when the blade is vertical. This cyclic twisting can quickly fatigue and crack the blade roots, hub and axle of the turbine.

Spalling (breaking up into fragments) of rotor shaft bearings can result in cracked rings, and in some cases, revolution of the inner rings around the shaft causes cracks in the shaft that can result in a total loss of the turbine. Figure 5 illustrates what can happen when the rotor bearings fail (in this case the blades landed 1 km away after having crossed a road).

Figure 5: The Loss of the Turbine Due to Rotar Bearing Flaws

6.4 Flaws on Transmission Bearings

The raceways of the roller bearings (usually self-aligning roller bearings) of the low-speed shaft can develop spalling over the entire circumference of the raceway, and excessively high temperatures can damage the bearings of the high-speed shaft. Raceway damage can negatively affect the transmission of forces to gears.

6.5 Flaws on Gears

Frequently, scuffing along the line of action as well as unsatisfactory tooth contact patterns are encountered. Highly loaded conditions result in chipping or micropitting.

6.6 Flaws on Generator Bearings

Generator bearing slippage and wear can ultimately cause inner rings to rotate around the shaft to the point that rotor makes metaltometal contact.

6.7 Flaws on Electrical Components

Flaws in the electrical system can also result in the outage of a wind turbine. Often damaged generator windings, short-circuit and overvoltage damage to controllers and electronic components as well as damage to transformers and wiring are found.

6.8 NDT Techniques In-Service

There are not that many NDT techniques used in-service, but it is well known that AE is able to locate and monitor both high damage regions and flaws in the blade structure.

7. Published Results from Research Studies in NDT

7.1 Acoustic Emission

A substantial amount of work has gone on in AE since 1993(3). A fatigue test of a wind turbine blade which was conducted at the National Renewable Energy Laboratory shows that fatigue tests of large FRP wind turbine blades can be monitored by AE techniques and that the monitoring can produce useful information(4). AE testing procedures, developed during a laboratory blade testing programme, have been applied to an in-service wind turbine blade in 2003(5). In the Framework of the Non-Nuclear Energy Programme, JOULE III from 1998 to 2002(6), the partners successfully developed a methodology for the structural integrity assessment of wind turbine blades, either in-service, or during certification testing, based on Acoustic Emission monitoring during static or fatigue loading. Within the framework of JOULE III, specialised pattern recognition software for AE data analysis and wind turbine automatic structural integrity assessment has been developed.

7.2 Full Scale Testing of Wind Turbine Blade to Failure

A 25 m wind turbine blade was tested to failure when subjected to a flapwise load[7]. With the test setup, it was possible to test the blade to failure at three different locations. The objective of these tests is to learn about how a wind turbine blade fails when exposed to a large flapwise load and how failures propagate. The report also shows results from the ultrasonic scanning of the surface of the blade and it is seen to be very useful for the detection of flaws, especially in the layer between the skin laminate and the load carrying main spar. AE was successfully used as sensor for the detection of damages in the blade during the test.

7.3 Wireless Detection of Internal Delamination Cracks in CFRP

In this study, a wireless system using a tiny oscillation circuit for detecting delamination of carbon/epoxy composites is proposed[8]. A tiny oscillation circuit is attached to the composite component. When delamination of the component occurs, electrical resistance changes, which causes a change in the oscillating frequency of the circuit. Since this system uses the composite structure itself as a sensor and the oscillating circuit is very small, it is applicable to rotating components. The wireless method is found to successfully detect embedded delamination, and to estimate the size of the delamination.

7.4 Structural Health Monitoring Techniques for Wind Turbine Blades

These experiments indicate the feasibility of using piezoceramic patches for excitation and a Scanning Laser Doppler Vibrometer or piezoceramic patches to measure vibration to detect damage[9]. Further testing of different smaller damages and types of damage is needed to verify the sensitivity of the methods. The resonant comparison method can be used for operational damage detection while the operational deflection shape method produces non-symmetric contours that are an easily interpretable way to detect damage in a structure that is not moving.

7.5 Infrared Thermography for Condition Monitoring of Composite Wind Turbine Blades

Infrared thermography has the potential for providing full-field non-contacting techniques for the inspection of wind turbine blades[10]. For application to turbine blades, the sensitivity of the thermal imaging has been shown to be suitable for non-destructive examination during fatigue testing; furthermore, it is thought that for blades *in situ*, the wind loading conditions may be sufficient to create effects detectable by thermal imaging.

8. Standards

Project certification can be done to a degree under the International Electrotechnical Commission (IEC)-CAP standard. However, the CAP standard is not sufficient as it stands. High quality and efficient standardisation and certification are vital given the number of turbine types. Standards designed for one market segment can be inappropriate in another, and standards across the segments should normally be limited to essential operating and safety standards.

The standards that are perceived to be lacking in some way need to be identified and appropriate actions for new standards and background research initiated. There is also the need to develop turbine type categories on the basis of ISO/IEC and CEN/Cenelec standards.

8.1 At Manufacture

At the time of manufacture, the inspection standards should be designed for detecting manufacturing flaws (which are often quite different from in-service flaws) and to ensure that the wind turbines are manufactured correctly.

8.2 International Standards Activities

The American Wind Energy Association (AWEA) is the recognised US industry organisation for standards development. AWEA maintains contact with the IEC standards development activities through the involvement of staff members and industry representatives on TC-88 standards subcommittees.

There are at present published standards on the safety requirements for large wind turbines, small wind turbine systems, acoustic emission measurement techniques and performance measurement techniques (see also *www.ansi.org*).

8.3 European Standards

The current state of European standards in relation to wind turbines is summarised in Table 1.

Table 1: European Standards

Project number	Standard reference	Technical body	Title (EN)	Directive
7125	EN 61400-2:1996	CLC/TC 88	Wind turbine generator systems – Part 2: Safety of small wind turbines	73/23/EEC
12538	EN 61400-1:2004	CLC/TC 88	Wind turbine generator systems – Part 1: Safety requirements	73/23/EEC
16198	prEN 61400-2:2005	CLC/TC 88	Wind turbine – Part 2: Design requirements for small wind turbines	73/23/EEC
11914	EN 50308:2004	CLC/TC 88	Wind turbines–Protective measures – Requirements for design, operation and maintenance	98/37/EC pending

Conclusions

There is a need to develop:

- European standards for use by developers, investors and insurance companies on risk, economic viability, performance, reliability, and owners and managers of wind farms.
- European certification and accreditation systems for components, turbines and projects, including standards for the NDT of components and turbines, particularly for in-service NDT.

Acknowledgements

The authors would like to thank Jacobi Consulting Ltd for funding this study and giving permission to publish this review. The full report can be obtained by written request to Jacobi Consulting Ltd.

(M A Drewry and G A Georgiou are with the NDE Group, QinetiQ, Farnborough, UK. The Authors can be reached at george@jacobiconsulting.co.uk and melody.drewry @gmail.com respectively).

References

1. *'Wind Energy – The Facts'*, Vol 3, 1.3.2, EWEA and the European Commission's Directorate General for Transport and Energy (DG TREN).
2. 'Wind turbines. Cause investigation and consulting services', Allianz Center for Technology AZT (Germany), 15 May 2003.
3. J Wei and J McCarty, 'Acoustic emission evaluation of composite wind turbine blades during fatigue testing', *Wind Engineering*, Vol 17, No 6, pp 266-274, 1993.
4. A G Beattie, 'Acoustic emission monitoring of a wind turbine blade during a fatigue test', Department 9742, Sandia National Laboratory (USA), 1997.
5. M J Blanch and A G Dutton, 'Acoustic emission monitoring of field tests of an operating wind turbine', Energy Research Unit, CLRC Rutherford Appleton Laboratory (UK), 2003.
6. AEGIS, 'Acoustic emission proof testing and damage assessment of wind turbine blades', [JOR3-CT98-0283], September 1998 to October 2002.
7. E R Jørgensen, K K Borum, M McGugan, C L Thomsen, F M Jensen, C P Debel and B F Sørensen, 'Full scale testing of wind turbine blade to failure – flapwise loading', Risø National Laboratory (Denmark), June 2004.
8. R Matsuzaki and A Todoroki, Tokyo Institute of Technology, 'Wireless detection of internal delamination cracks in CFRP laminates using oscillating frequency changes', *Composites Science and Technology* 66 (2006), 407-416, July 2005.
9. A Ghoshal, M J Sundaresan, M J Schulz and P F Pai, North Carolina A&T State University and University of Missouri, 'Structural health monitoring techniques for wind turbine blades', *Journal of Wind Engineering and Industrial Aerodynamics* 85, 309-324, 2000.
10. G M Smith, B R Clayton, A G Dutton and A D Irving, 'Infrared thermography for condition monitoring of composite wind turbine blades: Feasibility studies using cyclic loading tests', Wind Energy Conversion 1993, Proceedings of the 15th British Wind Energy Association Conference, York (UK), pp. 365-371. Edited by K F Pitcher, Mechanical Engineering Publications Ltd. ISBN 0-85298-901-6, 6-8 Oct 1993.

9

Offshore Wind Power on the Horizon

A New Energy Frontier for Oceans, People and Wildlife

Jeremy Firestone, Sandy Butterfield,
Christina Jarvis and Jack Clarke

This article discusses how offshore wind holds much promise as a means to meet electrical demand in coastal states, particularly if potential sociocultural and environmental impacts are thoughtfully considered and addressed. With several manufacturers developing large wind electric turbines designed to be placed offshore in waters up to 30 m in depth, offshore wind technology and its potential is discussed for offshore wind development off Cape Cod and Long Island. Discussion regarding the impacts and benefits of offshore wind development on humans and wildlife with factors underlying public opinion on offshore wind is done. Floating platform systems at some deep sites for economy and feasibility are a reality with possible environmental changes like habitat disruption and possible habitat enhancements.

Recently, it has been recognized that large wind resources exist above the Atlantic Ocean, close to east coast cities. Several manufacturers have developed large wind electric turbines designed to be placed offshore, in waters up to approximately 30m in depth. To date, they have been placed in European waters. The resources, the technology, and the economic viability have all come together in the eastern US, with large-scale deployment potentially beginning in 2012. A comprehensive regulatory framework is currently being developed as well [1-2]. This paper describes offshore wind technology and offshore wind potential; development of offshore wind proposals off of Cape Cod and Long Island; impacts and benefits of offshore wind development on humans and wildlife; and factors underlying public opinion on offshore wind.

The vision for large-scale offshore wind turbines is not new. Floating systems were suggested as early as 1972 [3], but it was not until the mid-1990s, after the commercial wind industry was well established, that the topic was taken up again by the mainstream research community and commercial interests. Europe has led with 25 offshore wind since 1991 totaling around 1100 MW [4] driven by land use constraints, climate change reduction efforts, governmental financial incentives/ mechanisms, and an abundance of offshore wind resources. Fixed-bottom technology has been deployed on "mono-piles" and "gravity foundations" in water depths less than 20m. These projects have demonstrated the technology for further commercial applications resulting in future offshore wind target development of 3.5GW by 2010 and 20-40 GW by 2020 [ibid]. As the technology looks to advance into deeper water, support structures will become larger and more costly, motivating designers to mount the largest turbine possible in an effort to minimize the total cost per megawatt. Ideas include floating platforms that may be economical for deploying offshore wind turbines at some deep sites.

The offshore wind resource is extremely abundant, with the US energy potential ranked second only to China [5]. Figure 1 shows a US wind resource map by water depth. While wind resources are available in water depths of less than 30m, deeper water areas provide significant additional resources. The resource estimates shown were calculated by the National Renewable Energy Laboratory [6]. The calculations include all regions with annual average winds above 7.5 m/s – class 5. All windy water areas are considered from the shore out to 50-nautical miles and no exclusions were considered in the final numbers presented. The energy conversion estimates are based on 5-MW per square km of installed capacity. The conversion efficiency assumptions are left for the reader to determine.

Figure 1: US Offshore Wind Energy Resource by Depth

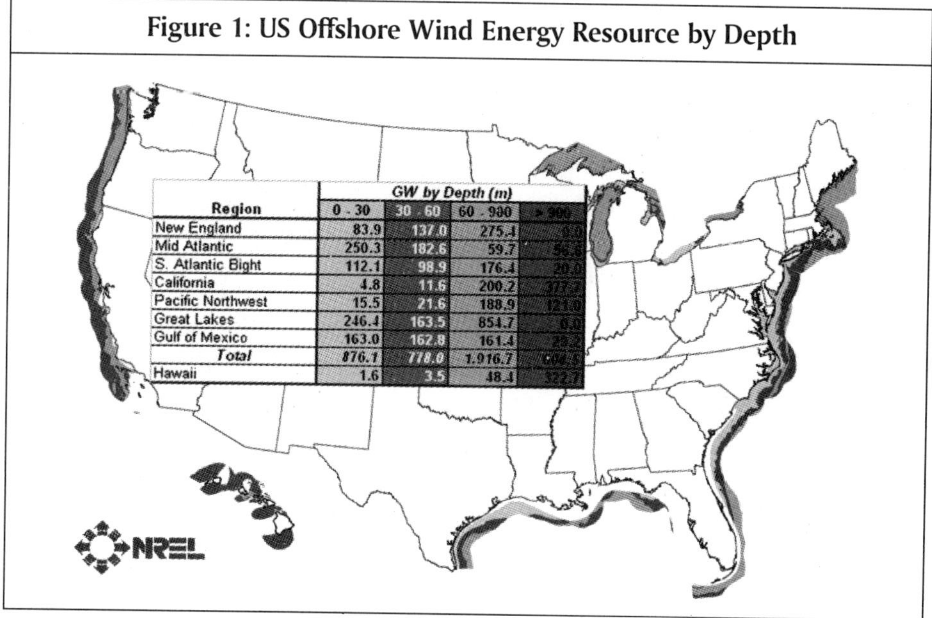

Region	GW by Depth (m)			
	0 - 30	30 - 60	60 - 900	> 900
New England	83.9	137.0	275.4	0.0
Mid Atlantic	250.3	182.6	59.7	56.6
S. Atlantic Bight	112.1	98.9	176.4	20.0
California	4.8	11.6	200.2	377.7
Pacific Northwest	15.5	21.6	188.9	121.0
Great Lakes	246.4	163.5	854.7	0.0
Gulf of Mexico	163.0	162.8	161.4	29.2
Total	*876.1*	*778.0*	*1,916.7*	*604.5*
Hawaii	1.6	3.5	48.4	322.7

The first two projects to file complete permit applications off the US coast are found in Nantucket Sound (bounded by Cape Cod, and by the inhabited islands of Martha's Vineyard and Nantucket) and off the Atlantic coast of Long Island, NY. The Nantucket Sound project is being developed by a private company, Cape Wind Associates. Its proposal calls for the installation of 130 turbines in 24 square mile area. The project will be 4.7 miles from the nearest shore location on Cape Cod and considerably further offshore from Martha's Vineyard and Nantucket. Cape Wind plans to install 3.6 megawatt (MW) turbines that would rise 420 feet from sea level to the tip of the blade. This development proposal is projected to generate the equivalent of about 3/4 the electrical needs of Cape Cod, or 10% of the demand of the entire state of Massachusetts [7]. The then lead federal agency, US Army Corps of Engineers (USACE), issued a Draft Environmental Impact Statement (DEIS) on the project in November 2004 [8]. Section 388 of the Energy Policy Act of 2005 provides for a transition of federal lead agency authority from USACE to the Minerals Management Service (MMS). MMS issued a revised DEIS in January 2008, with the final EIS expected before the end of 2008.

In contrast to the private-developer driven Cape Wind project, the project proposed to be established off the Long Island south coast is being developed by the Long Island Power Authority (LIPA), a state power authority, and FPL Energy, the largest wind energy developer in the United States. In early 2002, a LIPA study revealed an untapped potential of over 5,000 MW of wind energy off Long Island's shores [9]. A subsequent site assessment identified a 52-square nautical mile area off the south shore of Long Island for development consideration [10]. LIPA issued a Request for Proposals for a firm to develop, construct, own and maintain a 100 to 140 MW offshore wind park in the preferred siting area, with FPL Energy ultimately selected for Project development. The proposed Long Island Offshore Wind Park consists of 40 wind turbines, each capable of generating 3.6 MW of electricity. The project will be located southeast of Jones Beach State Park and southwest of Robert Moses State Park. The nearest turbine to shore will be approximately 3.6 miles while the farthest is approximately 5.5 miles. The turbines will be arranged in multiple rows, spaced one-third to one-half mile or more from each other. On April 26, 2005, LIPA and FPL Energy filed a joint Section 10 application with the USACE seeking authorization to install a 140 MW offshore wind energy park off the South Shore of Long Island. LIPA will be responsible installation and operation of the buried 138 kilovolt cable while FPL Energy will construct, own, operate and maintain the offshore substation and wind turbines. As part of the transmission line environmental review, LIPA will be submitting an Article VII application for approval of the 138-kilovolt transmission line under state jurisdiction. NEPA compliance is currently being addressed by the MMS.

Offshore wind projects, have numerous public benefits but are not controversy-free and indeed raise environmental and socio-cultural issues. Offshore wind projects must address impacts to local and migrating species of wildlife are frequently identified as an issue of public concern. Impacts to wildlife species may occur during all phases of offshore wind development, and may include habitat alteration, temporary or permanent habitat displacement, increased levels of underwater noise and vibration, and in some cases, mortality [11]. While avian and bat collisions with wind turbine rotor blades have frequently been recognized as an impact of terrestrial wind facilities, these may be reduced offshore as has been demonstrated in operating offshore wind farms in Europe. The nature, direction (positive or negative) and severity of marine mammal, sea turtle bird, fish and benthic impacts at offshore facility will depend on a variety of factors, including geographical and spatial characteristics of the site itself;

turbine design and lighting; and abundance, distribution, and population status of marine species [12-13]. Many of the wildlife impacts that occur during construction and decommissioning phases will likely be of short duration and localized to the immediate area, resulting in temporary habitat displacement for some species. Impacts that occur during the operation of an offshore wind facility, such as increased levels of underwater noise and vibration, represent a long-term change in habitat to which species must either adapt to or abandon the wind farm area to find new habitat. On the other hand, artificial reefs may provide habitat enhancements.

Wildlife impacts (and other externalities) that result from wind power generated electrical power, however, should not be viewed in isolation, but rather need to be considered *relative to* the wildlife impacts and other environmental consequences from other forms of energy production [8, 11-13]. Impacts from fossil fuel burning power plants, for example, include the entrainment and impingement of billions of aquatic organisms; thermal pollution of nearby waterbodies; habitat alteration from resource extraction, processing, and transportation activities; air pollutant emissions and greenhouse gases; and acidification of aquatic and terrestrial habitats [12]. These impacts are likely to have long-lasting implications for wildlife, such as habitat displacement, physical impairment, and reduced breeding potential. In sum, a complete and balanced evaluation of offshore wind energy must consider not only the wildlife impacts of wind development, but the wildlife impacts should the potential for offshore wind energy *not* be realized – that is, continued growth of fossil fuel power production and its associated environmental and wildlife degradation.

Turning last to socio-cultural concerns, the Cape Wind offshore wind proposal has generated a vocal, well-funded, and politically-connected opposition and support groups. The question of whether opposition and support is based on well-reasoned judgments about the proposed project's positive and negative impacts. An analysis of a survey by Firestone and Kempton of approximately 500 local residents provides some answers [14-15]. They found that the majority expects negative impacts from the project, with a much smaller percentage expecting positive effects. The most frequently mentioned factor identified as affecting one's position was marine life/environmental impacts, followed by electricity rates, aesthetics, and fishing and boating impacts [14]. When Firestone and Kempton compared survey respondents' expectations with the findings of scientific studies, they found that many of the population's beliefs appear to be inconsistent with those findings. They identify a need for better information on electricity rates; fishery, marine mammal and bird impacts; air quality improvements; and job creation.

Firestone and Kempton [14] found Cape Wind supporters to be younger, better educated, and more likely to own their own home, while opponents are more likely to earn over $200,000/year and more likely to expect to see the project from their daily routine. They found some support for a NIMBY hypothesis, but not for the stereotype that opposition is simply "rich coastal property owners." They also tested whether a change to the project would affect support (e.g., on land, further offshore, proposed by a local government, first of many). The biggest change was a net 36% increase in support if the Cape Wind project was the first of 300 such projects, in sum having proportionately larger impacts as well as greater benefits. They surmise an "important part of the opposition to offshore wind power projects is that the proponents have not always successfully articulated a larger vision—that offshore wind is abundant in many areas of the world, including this region, off the US East Coast, and that large scale development is a plausible outcome of individual successful projects" [14].

In sum, offshore wind holds much promise as a means to meet electrical demand in coastal states, particularly if potential socio-cultural and environmental impacts are thoughtfully considered and addressed. We are hopeful that Congress' recent action of authorizing MMS to regulate offshore renewable energy development is a significant step in that direction [1].

Fall 2008 Project Update

Given concerns over projects costs, LIPA put the Long Island Offshore Wind Project on hold in August 2007. However, in September 2008, LIPA announced plans to develop a 100 turbine wind-farm off of Queens, New York, with the utility, Con-Ed. Recent action in other coastal states in the mid-Atlantic also strongly suggests the emergence of a vibrant offshore wind power industry in the United States. Bluewater Wind, a subsidiary of the Australian firm, Babcock and Brown, won an all-source bidding contest in Delaware besting coal gasification and natural gas plant bids. In 2008, Bluewater signed the first long term power purchase agreement (a 25-year contract) to supply energy from an offshore wind power project to an electric distribution company in the Americas. Bluewater also agreed to supply wind power to nine Delaware municipalities. When constructed the wind farm will be between 230 MW (the amount under the agreements) to 600 MW, and will be located about 12 miles from the Delaware coastline. Two other states, New Jersey and Rhode Island, each issued requests for proposals for offshore wind power and separately announced Deepwater Wind (in partnership with a utility in New Jersey, and by itself in Rhode Island) as the winning bidder. New Jersey also announced a commitment to develop 3 GW of offshore wind power by 2020. The US Department of Energy's recently announced goal of 54 GW of offshore wind power by 2030 [16] and the American Wind Energy Association's first national offshore wind power workshop in the United States, which was attended by more than 450 individuals, are additional evidence that offshore wind power is ready to breakout in the Americas.

(Jeremy Firestone is Associate Professor, Marine Policy, University of Delaware College of Marine and Earth Studies. The author can be reached at jf@udel.edu

Sandy Butterfield is Chief Wind Engineer for the National Renewable Energy Laboratory US Energy Department. The author can be reached at Sandy_Butterfield @NREL.GOV

Christina Jarvis, University of Delaware College of Marine and Earth Studies

Jack Clarke is director of advocacy for Mass Audubon. The author can be reached at jclarke@massaudubon.org).

References/Endnotes

1. MMS, Alternative Energy and Alternative Uses of Existing Facilities on the Outer Continental Shelf, Proposed Rule, 73 Federal Register 39376-39504, July 9, 2008.
2. Firestone, J., W. Kempton, *et al.*, 2004. "Regulating Offshore Wind Power and Aquaculture: Messages from Land and Sea," *Cornell Journal of Law and Public Policy*, 14(1): 71-111.
3. Heronemus, W. E. "Pollution-Free Energy From Offshore Winds", 8th Annual Conference and Exposition Marine Technology Society, Washington D.C., September 11-13, 1972.

4. EWEA, Delivering Offshore Wind Power in Europe: Policy Recommendations for Large-scale Deployment of Offshore Wind Power in Europe by 2020, *http://www.ewea.org/fileadmin/ewea_documents/images/publications/offshore_report/ewea-offshore_report.pdf*
5. Musial, W.D.; Butterfield, C.P. "Future for Offshore Wind Energy in the United States" NREL/CP-500-36313 – EnergyOcean Proceedings, Palm Beach, FL, June 2004.
6. Musial, W.D. "Offshore Wind Electricity: A Viable Option for the Coastal United States", *Marine Technology Society Journal*, Fall 2007.
7. Cape Wind Associates. 2004. America's First Offshore Wind Farm on Nantucket Sound, *http://www.capewind.org*
8. US Army Corps of Engineers. 2004. Cape Wind DEIS, *http://www.nae.usace.army.mil/projects/ma/ccwf/windfarm.htm*
9. LIPA. 2002. Phase I Preliminary Wind Assessment, *www.lipower.org/pdfs/projects/wind/phase1.pdf*
10. LIPA, 2003. Phase II Siting Assessment, *www.lipower.org/pdfs/projects/wind/phase2.pdf*
11. Jarvis, C., 2005. Evaluation of the Wildlife Impacts of Offshore Wind Development Relative to Fossil Fuel Power Production. Master's Thesis, University of Delaware, *www.ocean.udel.edu/windpower/*
12. BLM. 2005. Final Programmatic EIS on Wind Energy Development.
13. European Commission. 2003. External Costs: Research Results on Socio-environmental damages due to electricity and transport.
14. Firestone, J. and W. Kempton. 2007 "Public Opinion about Large Offshore Wind Power: Underlying Factors," *Energy Policy*, 35: 1584-1598 (2007), available at *http://www.ocean.udel.edu/windpower/docs/FireKemp07-PubOpinUnderly.pdf*
15. Kempton, W., Firestone, J., *et al.*, 2005. The Offshore Wind Power Debate: Views from Cape Cod, *Coastal Management*, 33: 121-151.
16. US Department of Energy, 2008. 20% Wind Energy by 2030: Increasing Wind Energy's Contribution to US Electricity Supply, DOE/GO-102008-257 (July 2008), available at *http://www.nrel.gov/docs/fy08osti/41869.pdf.*

10

Wind Farm Development and Operation
A Case Study

Ton van de Wekken

This paper tracks how implementation of wind energy has changed dramatically over the last two decades from stand-alone to major wind farm development with turbine varying from a few to more than 25. Wind farms with established capacity of 50 MW are not uncommon and wind turbines in the range of 4.5 to 6 MW are already available on prototype. The development of onshore wind farms will require double investment costs as compared to onshore, not counting the high expenses to be incurred for sea foundation, sea cable and special sea vessels for transport and erection of equipment. Wind farms cannot be considered cost-effective on investment and ROI parameter; most wind energy projects need the perquisite of incentive for using renewable energy and not emitting carbon emission, to be profitable.

1. Introductory Notes

Nowadays most of the European countries are well acquainted with the phenomemon wind energy. In the eighties and nineties of last century mainly in Denmark and

Germany the knowledge on wind energy was concentrated including the manufacturing and siting of a large number of wind turbines. The last decade Germany is still at the top in yearly installed wind power, however, Spain is following at close distance. The last 5 to 10 years a large number of European governments developed policies on promoting renewable energy sources including wind energy. Prominent examples are United Kingdom, France, Italy, the Netherlands and Eastern European countries where wind energy is already present or incentives are defined to boost the increase of renewables.

During the last 20 to 25 years the implementation of wind energy has changed dramatically. The eighties and early nineties of last century was stand alone application of wind turbines with installed power between 100 and 500 kW quite common while the development of wind farms was rare. Due to several developments, ranging from local policies to economical considerations, currently most initiatives concern wind farm developments. The number of wind turbines in a wind farm varies from a few to more than 25.

From 2000, the installed power per wind turbine varies between approximately 750 kW and more than 3 MW. Wind farms with an installed capacity over 50 MW are not uncommon. Wind turbines in the range of 4.5 to 6 MW are available as prototypes and test specimen but not yet commercially explored.

Besides the development of onshore wind farms a tendency is noticeable to offshore siting of wind farms. Characteristic of offshore wind energy is the high installed power per wind turbine and initial investment costs which are almost doubled with respect to onshore siting. The additional expenses are caused by the high costs for offshore foundations, sea cable and special sea vessels for transport and erection of the wind turbines. Also the operational costs are at least the double compared to onshore siting. A main advantage of offshore is the low nuissance to the environment and the excellent wind resources leading to a high degree of utilisation.

The costs of onshore wind energy ranges from 55 to 100 EUR/MWh mainly depending on the wind resource. For most locations wind energy is not cost effective and incentives are a prerequiste to make a wind farm profitable.

Figure 1a: Overview of European Wind Atlas, No data of the Balkans and Eastern Europe is Available

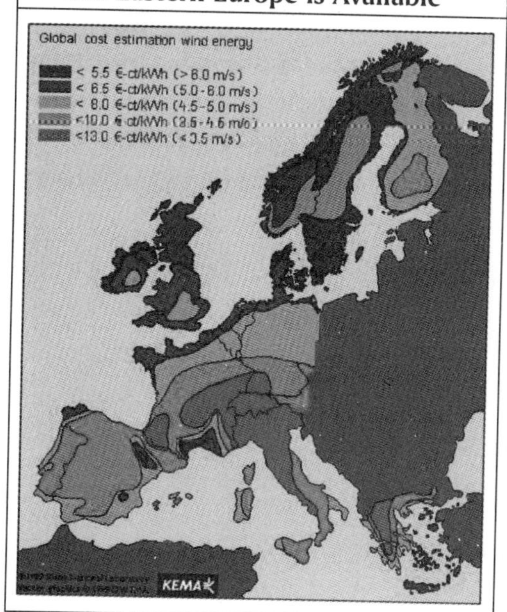

Figure 1b: Photograph of 10 MW Wind Farm on Flat Land Near the Atlantic Coastline

The European Wind Atlas demonstrates that Skandinavia, the UK, Ireland and the Atlantic coastline of the European continent have the best wind conditions for the developing of wind energy.

The time span from first initiative to final commissioning of a wind farm is subdivided into a development period and a building period. The duration of the building period is well known, for a small to medium size wind farm (< 15 MW) ¾ to 1 year and for a large wind farm 1 to 2 year. The technical life span of a wind turbine is 20 years.

The duration of the development period is less predictable. Mainly the time required for an irrevocable building permit is hard to predict. Depending on the mandatory procedure and the number and kind of objections put forward by concerned parties the required time span may vary from approximately ½ year to more than 5 years.

The financing of a wind farm is a matter of a special concern. In the past a stand alone wind turbine was part of the capital properties of a company. In such a case the wind turbine financing is comparable with the finance of other capital expenses.

Nowadays, for most wind farms a separate legal entity is established with limited equity capital and for the greater part founded on loans. It is clear that under such conditions the financer(s) will demand for extra financial guarantees. Generally the development and operation of a wind farm can be subdivided into the following four phases:

- Initiation and feasibility (concluded by go/no-go)
- Prebuilding (concluded by go/no-go)
- Building
- Operation and maintenance.

The following case has been selected for this note:

- Wind farm with 10 MW installed power
- Built up from five wind turbines of 2 MW
- Hub height 80 meter
- Rotor diameter 80 meter, 3 bladed rotor
- Octangular foundation, gross dimensions 18x18 meter
- Nacelle weight 100 ton, tower weight almost 200 ton
- Life span 20 years

The four phases of development will be discussed separately in the next paragraphs of this note.

2. Wind Farm Initiation and Feasibility Phase

Main subject during this phase is that one or more appropriate sites are selected for possible siting of a wind farm. Main characteristics are investigated like number of turbines, installed power and hub height. The feasibility study comprises an inventory and assessment of the main project risks like the presence of sufficient wind resources, sufficient grid capacity and verification with the municipality zoning plan. The phase is concluded with a go/no-go decision for the next process step.

2.1 Site Selection and Wind Assessment

In order to develop and construct an economical feasible wind farm an inevitable first step is to obtain one or more appropriate areas of satisfactory dimensions. Already for a medium size wind farm, e.g. 5 wind turbines of 2 MW, a substantial area is required. Depending on the rotor diameter the required mutual distance between the wind turbines is 300 to 500 meter and further to limit the nuisance or for safety reasons the distance to the nearest dwellings and company buildings is also at least 300 to 500 meter.

Next step, immediately following pre-selection of the area, is to assess the corresponding local long-term wind climate. Generally speaking, potential wind farm sites are preferably vacant areas at the flat land or on top of hilly areas. In all cases the sites should be characterized with high and recurrent wind resources.

Purpose of the initial screening study is to identify and evaluate factors that may lead to a definitive cross out of a pre-selected area.

In case the site screening does not identify any prohibitive limitations the feasibility study may proceed.

The financial feasibility is a prerequisite for the development of a wind energy project. The wind resource assessment is of outstanding importance for the estimation of the yearly energy yield and determines for the greater part the financial feasibility. The energy available in the wind is proportional with the wind speed to the third power. Based on local wind speed data of meteo stations can be determined a local wind atlas of the planned wind farm. It is necessary to use minimal one year of wind data to avoid fluctuations in wind speed during the seasons. The wind atlas is related to a roughness map of the area which is needed to determine the wind speed at a specific site and height. The estimated wind distribution results in a yearly energy yield representing the gross income of the wind farm.

The yearly energy yield is calculated by multiplying the wind turbine power curve with the wind distribution function at site:

$$E_{yearly\ yield}\,(kWh) = Ef(w_i).P(w_i) \qquad i = 1, n$$

With

f: wind distribution function (yearly hours per wind speed interval)

P: windturbine power curve (power output as function of wind speed)

wi: wind speed at interval or "bin"i, common interval size is 0.5 to 1 m/s

i: number of wind intervals "bins" between cut-in and cut-out wind speed; generally from 3 to 25 m/s

Figure 2a shows a power curve (PV) of a 2 MW wind turbine with optimal efficiency, i.e. without noise reduction measures that usually lead to less energy generation. Figure 2b shows the most commonly used wind speed distribution based on the statistical Weibull function with shape factor 2 and average wind speed of 7 m/s. Based on the PV curve from figure 2a and Weibull wind speed distribution with shape factor 2.0 the gross energy yield corresponding to 7 to 8.5 m/s is presented in Table 1.

Figure 2a: Power Curve of a 2 MW Wind Turbine (Without Noise Reduction Measures)

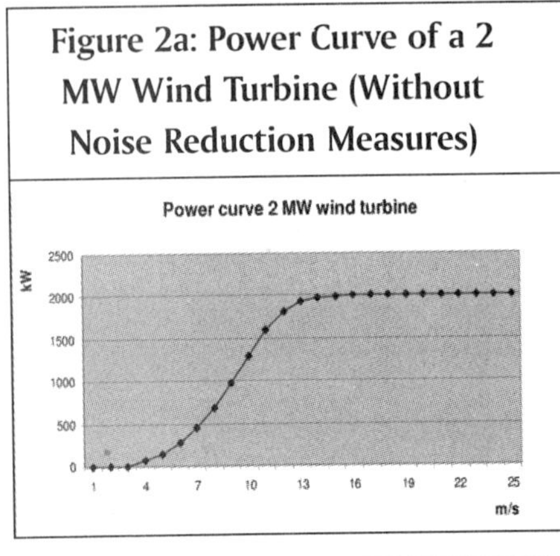

Figure 2b: Wind Speed Distribution, Average Wind Speed 7.0 m/s and Weibull Shape Factor k= 2.0

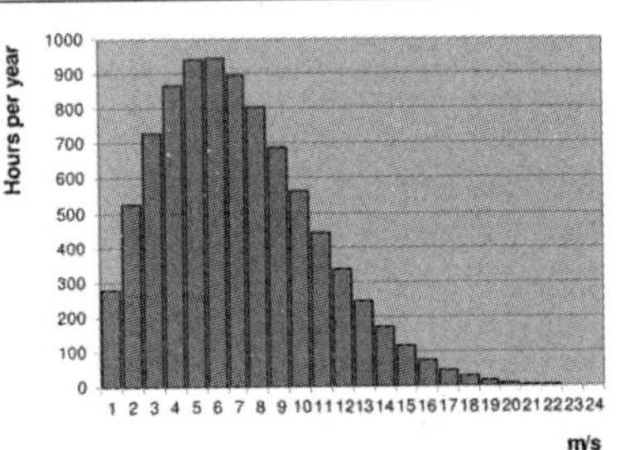

Table 1: Gross Yearly Energy Yield of 10 MW Wind Farm as Function of Average Wind Speed

Average wind speed	7.0 m/s	7.5 m/s	8.0 m/s	8.5 m/s
	Weibull shape factor k=2			
Wind Farm Power (MW)	10	10	10	10
Gross Annual Energy Yield (MWh)	27,000	31,000	34,000	37,000
Equivalent full load hours	2,700	3,100	3,400	3,700

Yearly gross energy yield of the 10 MW wind farm:

- 7.0 m/s average wind speed at hub height
- Wind speed according Weibull distribution function, shape factor 2.0
- No noise reduction measures required
- Gross energy yield 27,000 MWh
- Equivalent with 2,700 full load hours (utilization 0.31)

2.2 Technical Feasibility

Modern wind turbines are available in the power range from 0.75 to more than 3 MW having rotor diameters varying from 55 to more than 100 meter. Although in the past also two bladed rotors were used, nowadays only three blade rotors are commercially available. Generally, the hub height varies from 0.9 to 1.25 times the rotor diameter. Most manufacturers offer wind turbines with two or three different rotor diameters corresponding to low (large rotor), medium (standard rotor) and high or offshore (small rotor) wind climate.

At a first screening the outer dimensions of the available terrain are of importance. Wind turbines require a mutual spacing of at least four to five rotor diameters, corresponding with approximately 300 to 400 meter. A flat and undisturbed area is requested for buildings, trees and other obstacles lead to a lowering of the wind speed.

Next to the terrain orography it is well-advised to examine the local grid properties at an early phase. Main question is the distance to the nearest medium or high voltage substation with sufficient feed in capacity. It is advised to make already during the feasibility phase an appointment with the local grid operator to discuss grid connection including corresponding cost and planning.

2.3 Main Risk Assessment

In most countries it is prohibited that wind turbine rotors rotate above roads, railway tracks and water ways. A minimum distance has to be observed from wind turbines to public infrastructure. In northern countries and countries with continental climate specific attention has to be paid to probable icing problems. Ice developed on rotating rotor blades can be thrown far from the turbine and harm persons or result in material damage. Local authorities and concerned parties may demand for an aditional risk analyses in case in the vincinity of the planned wind farm at least on one of the following items count.

- Transport, storage or processing of hazardous goods
- Pipelines for transport of hazardous goods (also underground)
- Dwellings, company or public buildings
- Roads, railway tracks and water ways
- Medium and high voltage conductor
- Danger of icing.

2.4 Permitting Requirements of Local Authorities

The wind farm site has to meet national, regional and local requirements. In most countries special legislation is formulated for the environmental planning and the building aspects. During the wind farm planning phase, it is necessary to meet special zoning plan requirements. For instance it may be decided that, a specific zoning plan prohibit wind turbines or have laid down maximum heights for buildings. Under such conditions it has to be discussed with the local authorities possible ways and procedures to adopt the zoning plan to make the installation of a wind farm feasible.

In most European countries wind turbines have to be certified according to national or international safety standards especially developed for wind turbines. Manufacturers have to demonstrate approved cerification by a valid "type-certificate".

Planned area for the 10 MW wind farm:

- Check municipality zoning plan on competing activities and maximum building height.
- Mutual distance windturbines 400 meter
- Wind turbines in line, required length 1.600 meter
- Preferable within 300 to 500 meter from the wind turbines no dwellings and other buildings and further as less obstacles as possible
- Authorities or parties concerned may request a risk analysis if within 400 to 500 meter from the wind turbines other activities take place

2.5 Project Financing

As example is shown in the tables x and y the investment and operational cost of a wind farm of 10 MW. Wind turbines investment and operational costs are fully proportional with the installed capacity. Investment costs per MW installed are about 1.25 M€ and yearly operating cost somewhat more than 40 k€ per MW.

For a 10 MW wind farm, five wind turbines of 2 MW, the investment and refurbishment costs are estimated as follows:

Non recurring investment costs – price level 2006 –	Costs per wind turbine [K€]	Wind farm costs [K€]
Preparatory costs	100	500
5 wind turbines of 2 MW each	2.000	10.000
Wind farm civil and electrical infrastructure	200	1.000
Grid connection	200	1.000
TOTAL INVESTMENT (year 1)	**2.500**	**12.500**
Refurbishment (at year 10) - price level 2006 -	250	1.250
TOTAL	**2.750**	**13.750**

Yearly recurring operating and maintenance costs:

Yearly recurring operating costs – price level 2006 –	Costs per wind turbine [K€]	Wind farm costs [K€]
Wind turbine service, maintenance and insurance	50	250
Local taxes and contribution grid connection	10	50
Land lease	15	75
Daily management	5	25
Own electrity consumption	5	25
TOTAL	85	425

Investment and operational costs 10 MW wind farm:

- Initial investment costs M€ 12.5 or 1.25 M€/MW
- Yearly operational costs M€ 0.425 or 42.5 k€/MW
- Estimated refurbishment costs halfway the life span M€ 1.25; i.e. 10% of the initial investment

In most cases the wind farm is financed by a mix of own capital (equity) and bank loan. The amount of equity is mostly limited to 20 to 40% of total investment. It may be interesting for companies, investment groups and private person to invest in wind energy for possible tax reduction benefits. A number of national governments in Europe have incentives to promote electricity production by renewable sources. A wellknown incentive is tax reduction for investments in "green" energy sources.

3. Pre-Building Phase

During this phase all preparatory work is done needed to start the building phase. The wind farm developer has to apply for all necessary permits and a power purchase agreement (PPA) has to be settled for selling the produced wind energy. The contractors for delivery of the wind turbines and corresponding civil and electrical infrastructure have to be selected. And last but not least, the project financing has to be arranged.

3.1 Wind Farm Design Including Energy Yield Predictions

Based on the wind resource assessment the most promising wind farm locations are studied in more detail. The computer models WasP, Wind farmer and WindPRO are some of the most wellknown models to calculate the energy yield of planned wind farms. Not only wind speed and wind direction distribution are taken into account but also the terrain orography; for instance a steep slope in the terrain will cause higher winds at the hill top. Such details in the modelling make that wind turbines are optimal sited by assuring optimal exposure to the wind. Generally the lay-out is optimised for exposure from the prevailing wind direction.

The mutual distance between the wind turbines has to meet the requirements of the manufacturers. When siting the wind turbines too close this may result in a lowering of the electricity production. Another, more serious, consequence may be the damaging of primary structural parts caused by the wake of upwind sited wind turbines. The minimum distance depends on the siting with regard to the prevailing wind direction. When siting perpendicular to the prevailing wind direction the distance has to be at least four and otherwise minimal five times the rotor diameter.

The gross energy yield of the wind farm is dominated by the local wind distribution and the siting of the wind turbines. To calculate the net energy yield it is needed to determine the anticipated losses. The gross annual energy yield has to be adjusted for:

- Wake losses
- Grid losses
- Availibility

3.1.1 Wake Losses

The wind speed down stream the rotor, the so-called wake, of the turbine is lower compared to the undisturbed wind speed resulting in a somewhat reduced performance of down stream sited wind turbines. The wake is characterized by extra turbulence which may lead to premature damage of main structural components. It is common practice to estimate the wind farm wake losses in the range of 3 to 4% of the gross energy yield.

3.1.2 Grid Losses

Grid losses are defined as the electrical losses between wind turbine switchgear and public grid connection, i.e. the location of the accountable metering. Depending on the lay-out the electrical losses are in the range of 2 to 3% of the gross energy yield.

3.1.3 Availability

The availability of a wind turbine is defined as the time the wind turbine is in operation or ready for operation with external conditional, for instance too low wind or grid loss, preventing the system from energy generation. The technical availability of the turbine is 97% or higher. This figure is based on data of modern operational wind farms.

Yearly nett energy yield of the 10 MW wind farm:

- Gross energy yield 27,000 MWh
- 3% wake losses
- 3% grid losses
- 97% availability
- Nett energy yields 24,500 MWh.

3.2 Permitting Procedures

3.2.1 Environmental Permit

The wind farm must comply with all regulations pertaining to environmental permitting. For the environmental permit a site plan and various environmental studies, amongst others on plant and animal life, are necessary.

3.2.2 Noise

The noise impact of the wind turbines on the environment is one of the major permitting issues. Wind turbines produce noise, mostly caused by the rotor blades and drive train. The distance to nearby located dwellings has to be sufficient to assure that the noise level at the house front is below the statutory norm. Next to a noise assessment and visual impact study, most of the authorities demand safety and risk assessment studies and shadow casting examinations. Shadow flickering is caused by sunlight reflecting on non rotating blades or tower. Shadow casting is due to periodically – about once per second – interrupting the sunlight by the rotating blades. Both flickering and shadow casting on dwellings and offices can be very annoying for people in there. Shadow casting is not regulated by law.

3.2.3 Safety

It is not allowed for wind turbines to rotate above roads or railway-tracks. In case icing on the rotor blades and nacelle may cause danger for persons and material in the near environment, protective measures have to be taken. A well-known measure is rotor standstill during icing up and release for starting up after visual observation all ice has disappeared.

3.2.4 Building Permit

Besides an environmental permit also a building permit is necessary for the wind farm. As part of the application most authorities require information on the visual impact of the wind farm on the environment. Visualisations of the planned wind farm in the existing environment serve as input for this assessment. Another part of the application is a detailed description of the wind turbines including foundation, exact locations of the wind turbines in the terrain, overview of the civil and electrical infrastructure and a copy of the valid type-certificate.

3.3 Grid Connection

Usually not each individual wind turbine is connected to the public grid separately. For the wind farm an internal grid is designed and installed. The voltage of the internal grid is preferred at medium voltage level, between 10 to 20 kV, in order to limit the losses. In most cases the wind turbines are electrically connected in a loop to ensure redundancy. The wind turbine generators operate mostly below 1000 V and each wind turbine is equipped with a transformer to transform the power from low to medium voltage level. The wind turbine transformer is located in the nacelle or tower base or in special housing next to the tower.

It depends on the public grid voltage if a central wind farm transformer is required to transform the medium voltage to grid voltage or only a switchgear installation including accountable metering provisions.

In some occasions no public grid to the wind farm is available or the existing grid has insufficient capacity. If the grid has to be extended or reinforced for the wind farm only the costs have to be born fully by the wind farm developer.

3.4 Feed-in Contract

In all cases a power purchase agreement (PPA) or feed in contract is necessary. The feed in tariff is composed from a contribution for producing and delivery of electricity and in most cases increased with an extra allowance for generating renewable energy or corresponding carbon credits. Depending at one hand on the wind resources and on the other hand on the investment and operational cost a feed in tariff of at least 60 EUR/MWh is required and for an economical viable project a tariff of 80 to 90 EUR/MWh is needed.

3.5 Selection of Suppliers

Based on the scope of supply and own working procedures the wind farm developer has to decide for a public tendering procedure or to decide to send only a pre-selected number of wind turbine suppliers a request for tendering. All necessarily information for the tendering has to be included in the "tendering enquiry documentation" also abbreviated to TED.

The TED has to include at least the following bid information:

- Sufficient information on the project including permitting status and project time planning
- Requested information from bidders, e.g. financial, technical and operational information
- Time schedule, a.o. tender closure date and assigned period for additional information, questions to ask and site visit;
- Contractual issues
- Scope of supply

- Technical specifications
- Maintenance and repair conditions
- Insurance and warrantee agreements.

With regard to the scope of supply it may be decided that the contracter delivers a turnkey wind farm including wind turbines, foundations, access roads, wind farm grid and connection to the public grid. Another option is to subdivide the delivery in a number of subdeliveries with different contractors, for instance to separate wind turbine delivery and installation from the civil and electrical works. The latter may have cost advantages, however the project developer shall act as main contractor and is responsibe for the wind farm entirely. Turnkey delivery by one main contractor is mostly the well-advised option.

It shall be noted that the offered wind turbine is certified by a recognized body and possesses a type-certificate valid for the local wind climate and wind farm lay-out. In order to avoid any problems on warrantees the contractor has to state formally that the delivered wind turbines are "fit for purpose" for the site.

3.6 Project Financing

As said before, for most sites in Europe wind energy is not yet cost effective. To promote wind energy, incentives are essential. The most applied promotion measures are:

- Subsidies (governmental or local) on investments in renewable energy sources
- Tax benefits for investing in renewable energy sources
- Reduced interest tariffs on loans for renewables
- Subsidies on the production of renewable energy (increased feed in tariff)

In case the financing is based for the greater part on loans the financers may ask for additional securities to guarantee that the loan can be repaid. The following securities can be asked for:

- Power purchase agreement with settled minimum feed in tariff for the loan period
- Guarantee that in years with moderate wind supply the income is sufficient for interest and reapyment

- Warrantees on supplied components and wind farm performance (availability and power curve) for the loan period
- Machine breakdown and business interrupt insurrance.
- Service and maintenance contract.

In annex A cashflow calculations with two different feed in tariffs are carried out for the 10 MW. In the calculations the financing costs are not taken into account. All expenses and incomes are based on price level 2006 so the influence of inflation is implicitly included in case expenses and income increases yearly with equal ratio.

Cash flow calculations for the 10 MW wind farm based on 20 year life span and price level 2006		
Feed in tariff	**60 €/MWh**	**85 €/MWh**
Gross income	1.47 M€/yr	2.08 M€/yr
Recovery period	> 15 year	> 9 year
Nett present value over 20 year (NPV)	6.1 M€	17.7 M€
Internal rate of return (IRR)	4%	11%

The calculations show that a feed in tariff of 60 €/MWh is needed to pay all costs without generating profit. Presuming the project is fully financed by a bank loan the project generates just sufficient cash to pay yearly interest (4%) and repay the laon and yearly operating costs. At a tariff of 85 €/MWh the project generates value for the owner (7%) in case the yearly financing costs ask again 4%.

4. Building Phase

The building phase includes all activities from commencement of the works up to take over of the operational wind farm.

4.1 Overview of Building Process

Following selection of the contractors and financial close of the project the manufacturing and building process commences. For the required exceptional transport the contractor has to verify the accessibility of supply routes towards site. Also adjacent to each wind turbine location sufficient space has to available for storage of the main components, assembly of the rotor and placing of the building crane. An area of approximately 50x80 square meter suits most applications. A 2 MW wind turbine requires at least a 600 tons caterpillar crane for hoisting of the tower parts, nacelle and rotor. Manufacturing and assembly of the main components takes fully place in the factories of the wind turbine supplier.

The following assembled main components are shipped to site:

- Foundation anchor or tube
- Three or four tubular tower parts
- Ground controller and switchgear
- Wind farm SCADA system
- Transformer (in case of ground based)
- Fully assembled nacelle (including gearbox, generator, yaw mechanism, mechanical break, converter and if applicable the transformer))
- Hub and rotor blades.

For a small and medium size wind farm the time between purchase order and transport ready is 6 to 9 months. Meanwhile the wind farm civil, including access roads, wind turbine foundations and substation, and electrical infrastructure is built. The time required for rotor assembly and constructing of the main structure takes two to three working days per wind turbine. Subsequently it takes 7 to 10 working days to finish the installation works and connecting to the grid. From the time all material is at site, the building time of a small and medium size wind farm takes only 2 to 3 month.

4.2 Quality Control during Production and Construction

It is common use that the contractor assigns a number of so-called "hold and witness" points for the client. These hold and witness moments are meant for the wind farm owner to audit the progress and quality of work including the verification that the components are in conformity with the specifications. Hold and witness moments are mostly planned immediately following a project milestone, for instance a main component ready for transport to site. Mostly hold and witness moments are link with instalments. The following hold and witness points are commonly used:

- Start of component production including audit of contractors quality system
- Factory acceptance test (FAT) of components ready for shipment

- Site acceptance test (SAT) of components delivered at site
- Several inspections during building at site, connected to milestones.

4.3 Commissioning and Take-Over

Following completion of the building and installation period and before take-over of the wind farm an overall inspection and commissioning of the works is carried out. Commissioning inspections are performed by representatives of contracter and coming owner. The commissioning may comprehend an elaborate testing and measuring plan but is mainly meant to verify proper installation and functioning of the installations. Also is verified that the delivery is complete without the omission of more or less important details. Normally in cooperation between the parties involved a check-list is formulated for the commissioning procedure. It is quite common that the first commissioning inspection results in a "punch list" with residual items. It depends on the gravity of the questions and further items on the "punch-list" if formal take-over may take place immediately after the first commissioning inspection or has to be delayed to approved second commissioning. Approved commissioning and take-over is related to last project payments.

5. Operation and Maintenance

Starting from date of take-over the owner is responsible for daily operation of the wind farm. Also from that date warrantee and maintenance contracts become valid. The technical and economical life span of a wind farm is anticipated as 20 years.

5.1 Daily Operation

Wind turbines are designed to operate unmanned. For normal operation no operator has to be available at site. It is common practice that medium and large size wind farms are equipped with a wind farm control and monitoring system (SCADA). By means of modem or Internet remote access to the SCADA system is available. The SCADA system provides reports on energy yield, availability and failure statistics. Main function of the daily operator is to verify regurarly that the wind farm is in optimal condition and performing according expectation. Also the operator is responsible that maintenance and repairs are carried out accordance to contract and within reasonable time.

5.2 Warrantees and Insurance

The following warrantees are common for the first five years following take-over:

- On delivered goods, including repairs and modifications
- Availability of individual wind turbines and wind farm, values of 95% or higher are not uncommon
- Warrantee on performance, for the exact wind supply is not to be predicted for a given year the warrantee is given on the power curve. Warrantees of 95% of the certified PV-curve of the wind turbine are common.

In case the availability or performance is below the warranteed value the difference between actual and warrantee values has to be settled by the supplier. Some suppliers offer warrantees up to 8 to 12 years or at least for a period comparable to the financing period. Insurances are required for:

- Third party liability
- Machine breakdown (e.g. material flaws, lightning strokes, fire, vandalism, fault by maintenance engineer and/or operator, etc.)
- Business interrupt (compensation for non production days following a machine breakdown event).

5.3 Maintenance and Repairs

Modern wind turbines require twice a year a preventive maintenance service. For a wind turbine in the MW-segment a planned preventive maintenance overhaul requires 2 tot 3 working days for two engineers. The work comprehends amongst others inspection and testing of the control and safety devices, repair of small defects, replacement or filling up of consumables like bearing grease and gearbox lubrication. The gearbox is the most vulnerable component and therefore subject of special interest during maintenance. At regular intervals oil samples are taken and investigated on pollution, filters are replaced and gearings are inspected on damages. The number of repairs differ largly between individual wind turbines and wind farms. The average yearly number of corrective actions per wind turbine is 3 to 4. Only those corrective actions are counted that need a visit on site of a service engineer. The mean down time per failure is 2 to 4 days. The cause of failure is equally divided between mechanical and electrical problems. Although not formally admitted by the manufacturers it is common practice that after 10 to 12 years of operation wind turbines need a major overhaul. The overhaul comprehends cleaning and repair work of the rotor blades and refurbishment of the drive train, i.e., replacement of bearings and if necessary replacement of gearbox parts.

ANNEXURE A

6. Examples of Two Cash Flow Calculations for a 10 MW Wind Farm

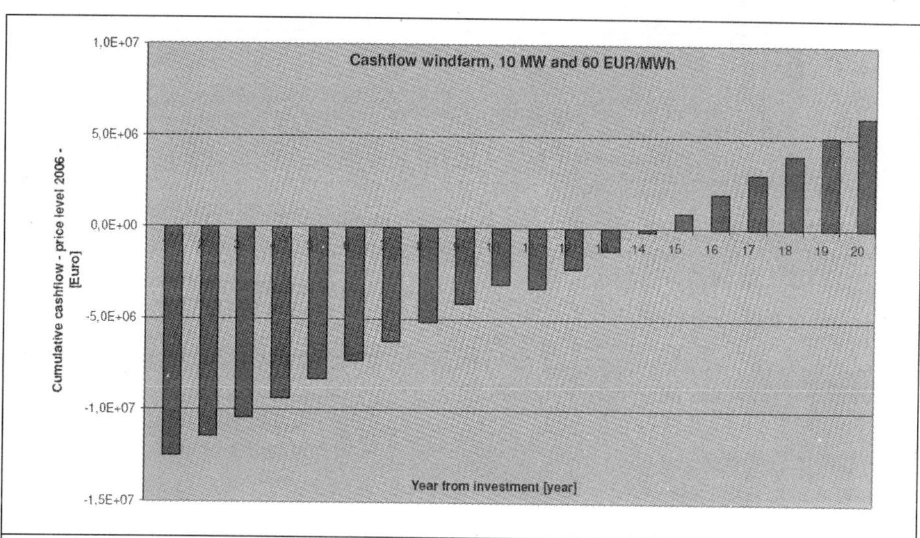

Summary of data scenario 1: wind farm of 10 MW	Low feed in tariff
No financing costs included!	
Price level 2006	
Investment costs (year 1)	• 12.500.000
Refurbishment costs (year 10)	• 1.250.000
kWh feed in tariff	• 0,060
Yearly wind farm production in kWh	24.500.000
Gross income from energy production	• 1.470.000
Yearly recurring costs	• 425.000
Recovering period	15 year
Net Present Value (price level 2006, 20 jaar)	• 6.100.000
Internal Rate of Return	4 %

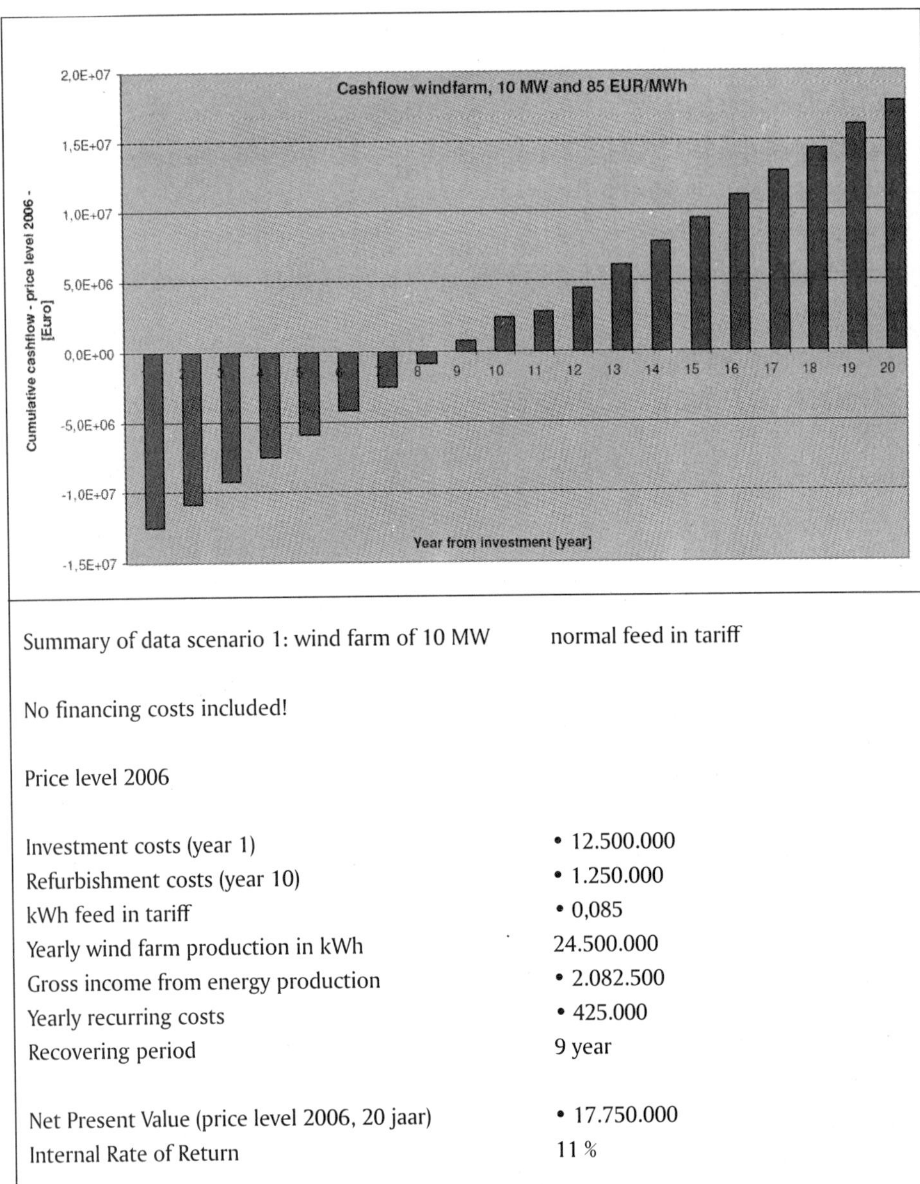

Summary of data scenario 1: wind farm of 10 MW normal feed in tariff

No financing costs included!

Price level 2006

Investment costs (year 1)	• 12.500.000
Refurbishment costs (year 10)	• 1.250.000
kWh feed in tariff	• 0,085
Yearly wind farm production in kWh	24.500.000
Gross income from energy production	• 2.082.500
Yearly recurring costs	• 425.000
Recovering period	9 year
Net Present Value (price level 2006, 20 jaar)	• 17.750.000
Internal Rate of Return	11 %

(Ton van de Wekken is from KEMA Consulting, founded in 1927; it is a commercial enterprise, specializing in high-grade business and technical consultancy, inspections and measurements, testing and certification. The author can be reached at Ton.vanderWekken @kema.com).

11

Impact of TradeWind Offshore Wind Power Capacity Scenarios on Power Flows in the European HV Network

J O G Tande, M Korpås, L Warland, K Uhlen and F Van Hulle

Simulations have been carried out for studying the effect of offshore wind on power flows in the European high voltage network. The simulations have been prepared applying a detailed model of the European power system for year 2005 and for 2020 with 152 GW onshore wind capacity and 48 GW offshore wind capacity. The analysis shows that bottlenecks between countries exist, but are generally reduced from 2005 to 2020. Both assumed grid development and wind power contribute to reducing the bottleneck costs for the 2020 scenarios. A modest cost can be attributed to wind variations. Actual bottleneck experienced costs by system operators are expected mainly due to market constraints for trade between countries, and due to country internal limitations in transfer capacity.

Source: http://www.tradewind.eu/fileadmin/documents/Papers/TradeWind_paper_7th_Windintegration_Workshop_Madrid.pdf

I. Introduction

TradeWind, a project initiated by EWEA (*www.tradewind.eu*) [1] investigates the necessary interconnection upgrade and electricity market rules to facilitate the integration of the future expected amounts of wind power. Scenarios of expected wind power capacity for the target years 2015, 2020 and 2030 distributed all over EU27 are the basis for the calculation of the wind power injections in the European grid and for studying the impact on the crossborder flows. This will form the quantitative basis for the formulation of the recommendations on market and interconnection.

Wind power is often said to contribute to congestions in the network, especially when the penetration in the power system increases. This paper, using the TradeWind approach and tools with respect to wind power scenarios and power flow simulation, analyses the impact of wind power – both onshore and offshore – on bottleneck costs in the European grid for a scenario when wind power will cover up to 10% of the EU electricity demand. It will be shown that – assuming a perfect market – wind power has a rather beneficial effect on the bottleneck costs in the EU grid.

The paper will first briefly describe the power flow model used for the TradeWind simulations. Then a description will be given of the wind power scenarios. In Section IV a description is given of the assumptions for the specific simulations performed to calculate the power flows and to analyse the bottleneck costs. The results are discussed in Section V.

II. Power Flow Model

The Power System Simulation Tool (PSST), developed by SINTEF Energy Research, is used for simulating the European grid. The structure of the simulation tool is shown in Figure 1. The inputs to the program are the grid layout, time series for load, time series for wind, generation capacity forecast for all generator types and generation costs for all generator types. For each hour the program updates the load, wind production and marginal cost of hydro units and runs a DC optimal power flow, which determines the power output of all generators and the power flow on all lines. For the optimal power flow calculation, the toolbox uses an external simplex LP solver, Clp [18], from Coin (COmputational INfrastructure for Operations Research) with a modified mex [17] interface to the C++ library.

Figure 1: The Main Simulation Structure of PSST

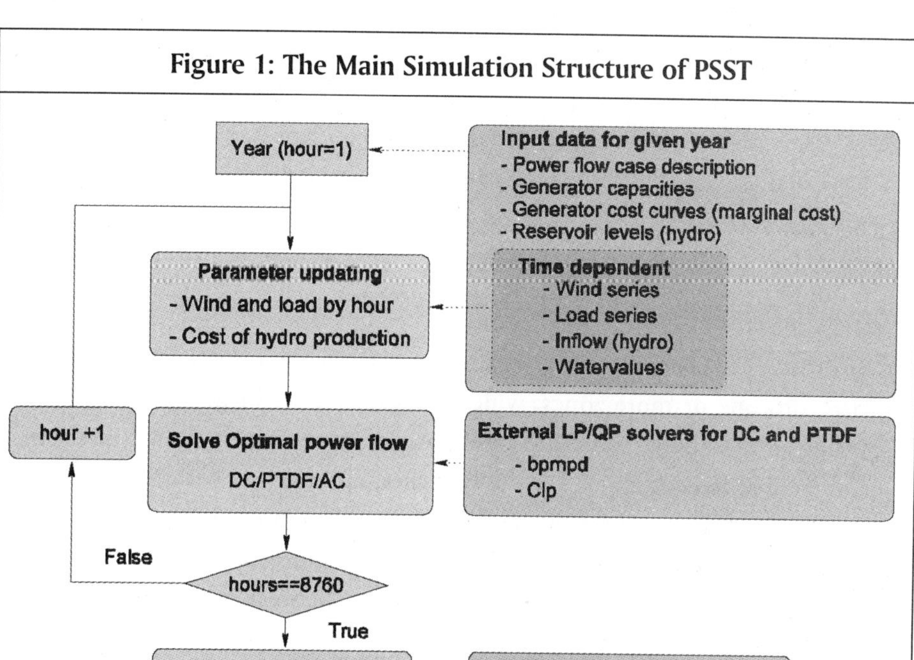

The free (controllable) variables in the optimal power flow problem are the power output of all generators and the flow on HVDC interconnections. The power outputs of the generators are dependent on the maximum and minimum capacity, the marginal cost relative to other generators and limitations of power flow on lines. Aggregated wind farms are modeled as generators with maximum power equal to the available wind power for the specific hour. The marginal cost is set to zero, so that wind power plants always will produce if not limited by grid constraints.

The European grid model that is used in the simulations consists of separate DC power flow data files for the Union for the Coordination of Transmission of Electricity (UCTE), Organization for the Nordic Transmission System Operators (Nordel) and Great Britain plus the island of Ireland. The power flow data for the three systems are merged together, making it possible to run an optimal power flow for the whole system. The base model consist of 1380 nodes, 2220 branches, 525 generators + wind farms and HVDC connections.

The UCTE network model is based on the approximated UCTE network stadium 2002 created by the team of prof. Janusz Bialek of University of Edinburgh [3], which covers the former first UCTE synchronous zone (i.e. excludes the Balkan states, Greece, etc). In the Tradewind project, grid reinforcements since 2002 and an approximate model of the former second UCTE synchronous zone have been added. See ref. [2] for further description of the UCTE network model, and the submodels for Nordel and Great Britain + the island of Ireland.

In order to study the main power flows in the system, each country has been divided into one or more zones, which consist of a set of generators and nodes according to their geographical location. A grid zone map is shown in Figure 2. Due to limited data available for the branches, all branches within a country has unlimited capacity. Only branches that connect two countries have a finite capacity. In addition, the Net Transfer Capacities by ETSO (European Transmission System Operators) are set as constraints for the power flow between countries.

Figure 2: Grid Zone Map

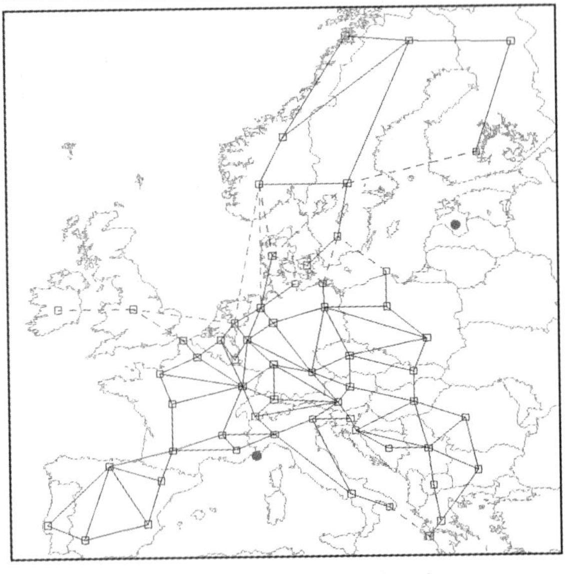

Each box refers to a grid zone and the lines show connections between zones. Dotted lines are HVDC connections. Red lines refer to connections that are added in the 2020 scenario.

Note: To view red lines indicated in the Figure readers may visit *http://www.tradewind.eu/fileadmin/documents/Papers/TradeWind_paper_7th_Windintegration_Workshop_Madrid.pdf*

III. Wind Power Scenarios

Wind speed data from the Reanalysis global weather model, combined with regional wind power curves and wind speed adjustment factors, is used for constructing synthetic wind power time series for the different grid model zones. The preparation of the wind speed data is based on RDPmapping (Reanalysis Data Points) as shown in Figure 3. The Reanalysis model, RDPmapping and different regional power curves are described in more detail in [4]-[6].

Figure 3: Reanalysis Data Points Used for Geographical Distribution of Wind Speed Input in the TradeWind Modeling

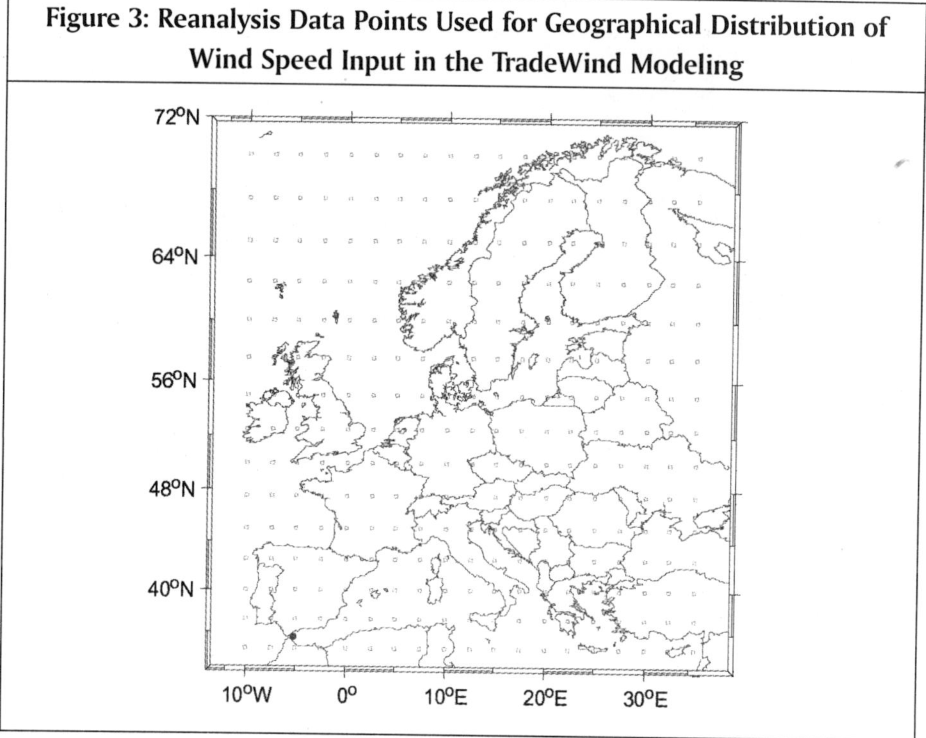

The scenarios for offshore wind power capacity are based on (a) an inventory in the EU27 within Work Package 2 of TradeWind and (b) a geographical mapping of offshore wind power developments in the EU. The first campaign of data collection (a) resulted in a set of capacity data attributed to the TradeWind grid model zones in each country. These data have been collected in the phase 1 of TradeWind early 2007 [4]. The data include low, medium and high estimations of installed wind capacity for the target years 2008, 2015, 2020 and 2030. The data have been provided by various national associations, industry and government contacts. An overview of the total installed offshore capacity for 2020 is presented in Figure 4. The resulting total offshore capacity and its evolution in time are presented in Table I and Figure 5.

Figure 4: Installed Offshore Wind Power Capacity, 2020 Scenario (Low, Medium and High)

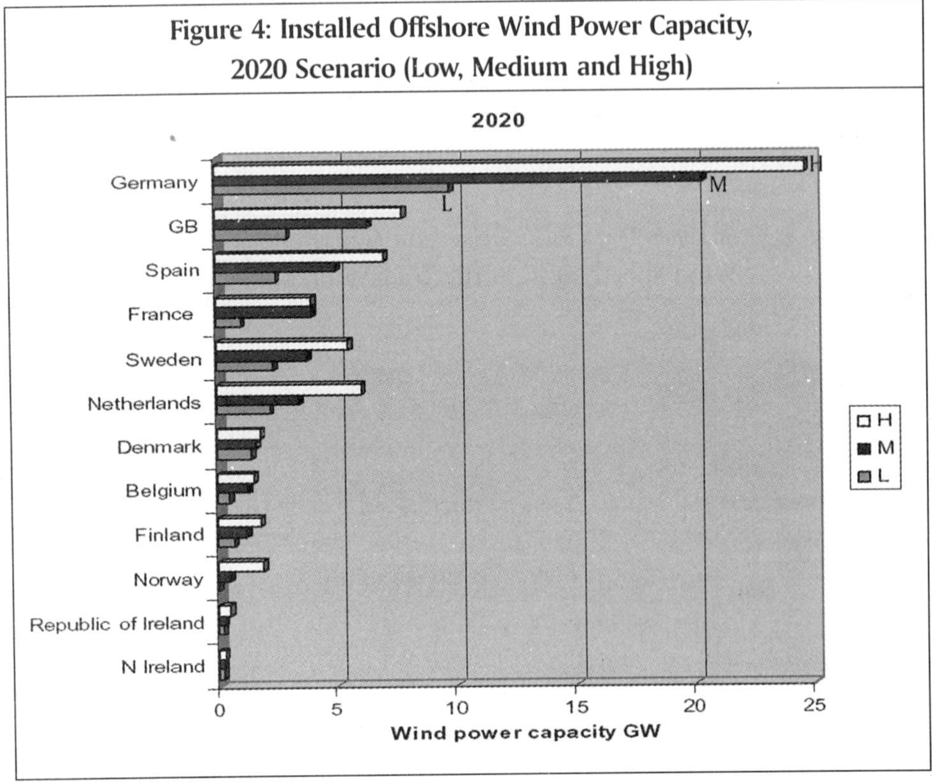

Table 1: Tradewind Offshore Wind Power Capacity Scenarios (GW)

Year	Low	Medium	High	EWEA Targets
2010	3.6	6.6	9.1	3.5
2015	16.0	25.0	36.2	12.0
2020	23.7	47.9	62.7	35.0
2030	40.2	98.6	134.3	120.0

Recently published targets of EWEA [7] are included in Table I for comparison. The TradeWind values corresponding to scenario 2015M and 2020M are higher, which shows that the development in time in these simulation scenarios is faster than assumed in [7]. It has to be remarked that the TradeWind scenarios have mainly a technical purpose, the values being used in order to evaluate the sensitivity of the power flows. The ultimate aim of the simulations is to identify significant grid situations relevant for grid extension and reinforcement planning.

Figure 5: Scenarios of Offshore Wind Power Capacity for TradeWind Simulations

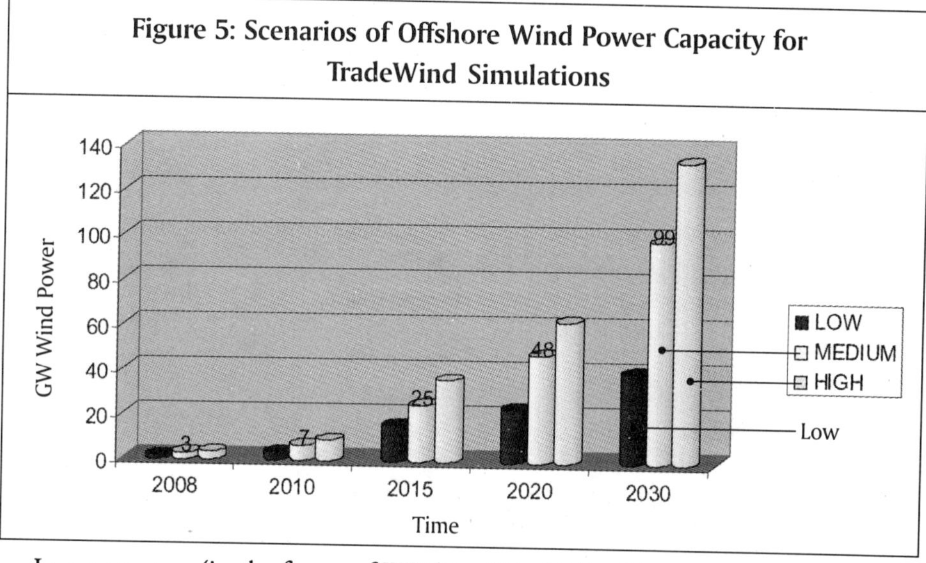

In a next step, (in the frame of WP6.2 of TradeWind), a geographical mapping has been made of future expected offshore wind power capacity based on individual projects and using different sources (market analysis, national 3 contacts) [8].

The purpose of this exercise is to serve as a basis for grid extension scenarios. The mapping has been carried out for a smaller number of countries than in the first step: only the developments in countries affecting the construction of transnational offshore grids in Northern Europe have been considered. The resulting basic data set for the mapping includes up to 125 GW wind power projects (reference year 2030). Every project is characterized by an identifier, country, rated capacity, geographical position coordinates of the wind farm centre point and a target date for start of production. A consistency check with the data collected in the first phase shows in general good agreements between both exercises for the total capacity in the countries considered. When looking at individual countries, there are some differences with respect to the data set in the first step, mainly due to recent revisions of targets in Netherlands, Germany and UK.

From this data set, a snapshot corresponding to 48 GW total installed offshore capacity has been extracted for further analysis in this paper. Indicatively, this 48

GW scenario corresponds to a medium scenario in 2020. The corresponding installed WP capacities in the countries are presented in Figure 6.

Figure 6: Offshore Wind Scenario 2020 for Northern Europe (Capacities in GW) and Location of Offshore Wind Plants

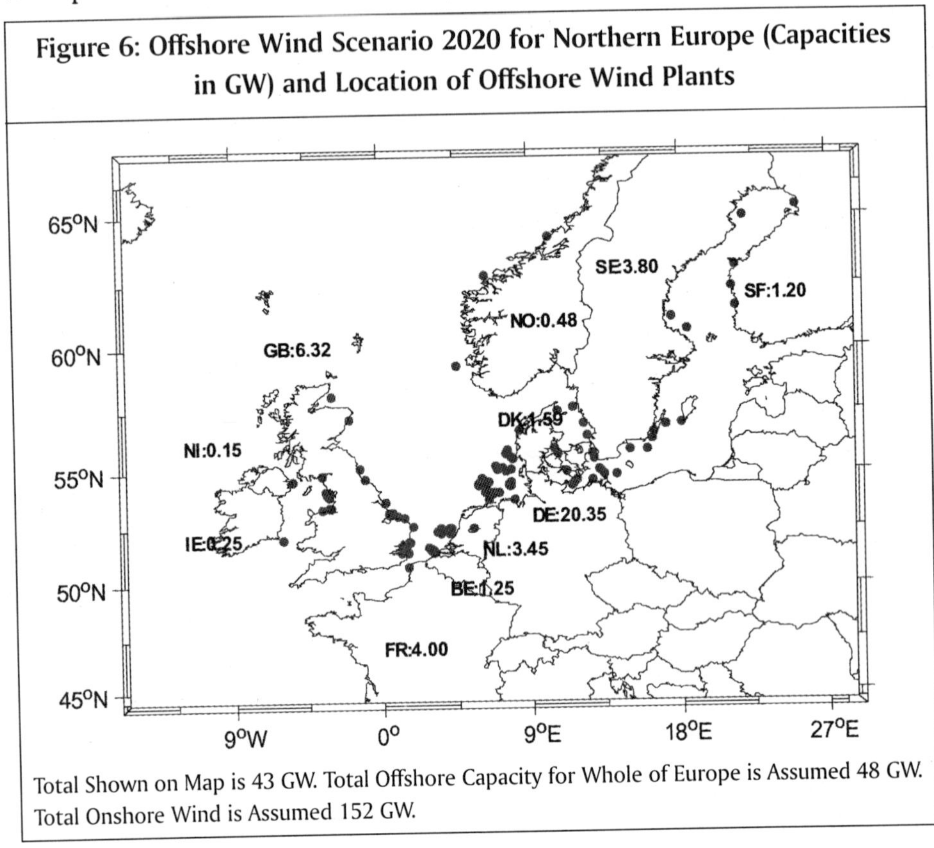

Total Shown on Map is 43 GW. Total Offshore Capacity for Whole of Europe is Assumed 48 GW. Total Onshore Wind is Assumed 152 GW.

IV. Simulation Set-up

The objective of this study is to analyse the wind power impact on bottleneck costs in the European grid. In order to do this, simulations have been run for year 2005 (base case) and year 2020 (onshore and offshore wind scenario). Bottleneck cost is defined as the difference in total generation costs between a simulation with full grid model and a simulation with a "copperplate model" (no constraints on power flow), divided by the total demand, or alternatively the total wind generation.

The offshore wind in this study assumed connected directly to the nearest onshore grid point. This is done for setting a starting point only, and should at a later stage be complemented by an analysis considering a dedicated transnational offshore grid for connecting the wind farms. The total installed generation capacities of the modeled generator types are shown in Table II. The total consumption in the 2005 and 2020 simulations are 3289 TWh and 4128 TWh respectively. The data for countrywise consumption and generation capacity (except wind) are based on official European statistics and scenarios [9]-[12].

Table II: Installed Generation Capacity for 3 Scenarios

Power Plant Type	2005 (GW)	2020 Onshore (GW)	2020 Offshore (GW)
Onshore wind	41,15	151,80	151,80
Offshore wind	0,85	0,00	48,12
Hydro	184,07	202,57	202,57
Nuclear	136,51	122,41	122,41
Lignite	71,97	77,84	77,84
Hard coal	113,52	96,49	96,49
Gas	126,37	251,74	251,74
Oil	27,80	19,78	19,78
Mixed oil/gas	39,71	40,14	40,14
Non attributable	10,93	20,66	20,66
Not clearly identifiable	2,59	1,32	1,32
Renewables (excl. hydro, wind)	14,30	38,84	38,84
Total capacity	769,78	1023,60	1071,70

Generation cost scenarios used in Tradewind are shown in Table III. These are mainly based on the DG TREN numbers [13], with additional use of refs. [14]-[16] in order to make cost scenarios for all the modeled generator types. The marginal cost of hydro power is modeled as a function of reservoir level, see [2] for details.

The bottleneck cost of wind power is estimated in this manner:

- A simulation of the full European grid, including wind power, is run for a full year (8760 hours), and the annual generation cost is divided by the annual load to get the average generation cost in €/MWh.
- Then the same simulation is performed using a copperplate model (i.e., all grid constraints removed). The difference in average generation cost between

the full grid model and the copperplate model is the bottleneck cost of system A (including wind).

- We now remove all wind power from the system and run the same simulations again to calculate the bottleneck cost of system B (excluding wind)

- The difference in bottleneck cost between A and B is the wind power contribution to the bottleneck cost. Note that this value can be both positive (wind power increases bottleneck cost) and negative (wind power reduced bottleneck cost).

In addition, the bottleneck cost of the wind power variations has been estimated. This is done using the same method as above, except that system B is replaced with a system where all wind plants have constant power output equal to the average wind power output (system C).

Table III: Marginal Cost Assumptions Using a CO_2 Price of 5 €/tonne for 2005 and 22 €/tonne for 2020

Power Plant Type	Marginal Cost [€/MWh]	
	2005	2020
Nuclear	11	11
Lignite coal	29.6	45.5
Hard coal	30.2	41.2
Gas	46.1	55.2
Oil	98.5	97.1
Mixed oil / gas	101.8	102.2
Non attributable	109.3	107.9
Not clearly identifiable	113.4	114.3
Renewables excl. hydro and wind	73.4	52.1

V. Results

The resulting generation and demand for the three scenarios considered are presented in Table IV. Wind supplies 2.5% of the annual load in 2005, 7.4% in 2020 onshore and 10.3% in 2020 offshore. No wind is constrained in 2005 and 2020 onshore, whereas 1% (5 TWh) of the potential wind generation is constrained for the 2020 offshore scenario. The constraint of offshore wind is likely due to the assumption of direct connection to shore, and would likely be zero had a dedicated transnational offshore grid been assumed. Offshore wind mainly contributes to reduction in generation from coal, oil and gas.

Table IV: Annual Demand and Generation (TWH)

	2005	2020 Onshore	2020 Offshore
Demand	3289	4128	4128
Potential wind	82	307	430
Actual wind	82	307	425
Constrained wind	0	0	5
Hydro	482	556	554
Nuclear	984	892	887
Lignite coal	489	539	503
Hard coal	789	822	816
Gas	436	831	765
Oil	1	~0	~0
Mixed oil / gas	7	3	3
Not clearly identifiable	11	11	11
Renewables (excl. hydro, wind)	6	162	160

The 2020 offshore scenario gives less wind generation than could be expected, i.e. above 3500 full load hours (FLH) for offshore wind farms could be expected on the average whereas the simulation gives only 2458 FLH. This is due to the assumed wind data (Reanalysis) that are good for giving the overall wind trends over Europe, but seemingly underestimate the wind conditions offshore. An adjustment of the assumed wind data is thus relevant, but out of scope of this paper.

Figure 7 shows duration curves of load, wind and net load (i.e., load – wind) for the 2020 offshore scenario. The curves show:

1. The total wind variations are smoothed thanks to the geographical distribution of the wind farms. For the assumed location of wind farms (with a significant concentration in Germany) the sum wind generation never exceeds more than 125 GW (62% of the installed capacity). It is a small probability (about 1%) that the sum wind generation goes down to 10 GW (5% of the installed

capacity). A more even geographical distribution of the wind farms would provide additional smoothing.

2. The total load varies between 295 and 660 GW, and the net load between 270 and 600 GW (with 99% probability). This means that although wind varies, it goes well together with load variations, and generally contributes to reducing the probability of peak load occurrence (and loss of load). Hence, wind may be assigned a capacity credit.

Figure 7: Duration Curves of Load (Upper), Wind Generation (Lower) and Net Load (Load – Wind) for 2020 Offshore

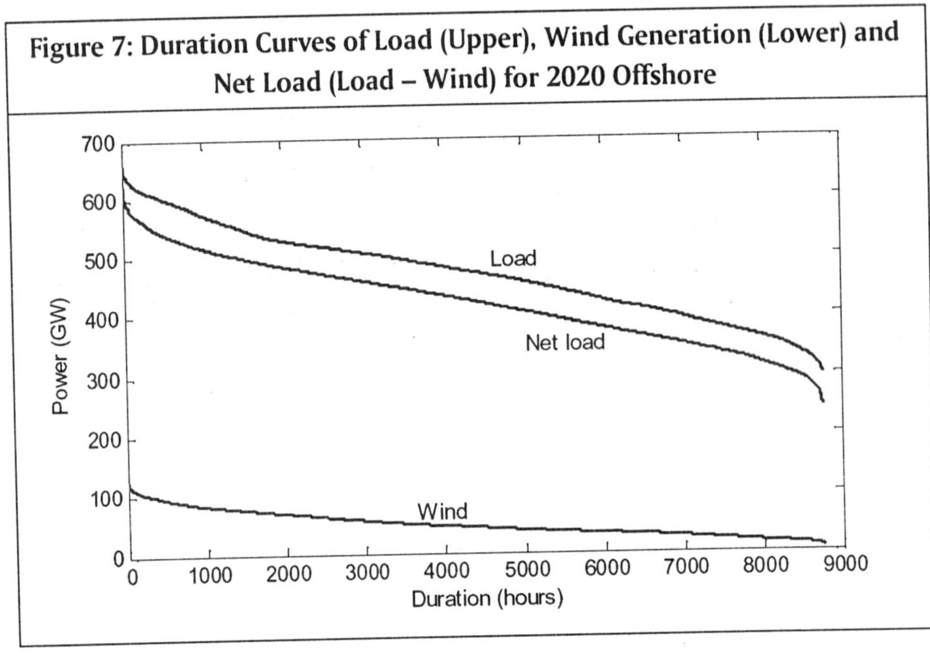

Table V shows the bottleneck costs resulting from the simulations, i.e. considering only cost of constrained transmission between countries and assuming perfect market conditions. The costs of these are generally reduced from 2005 to 2020 by the assumed grid development, but also wind contributes to reducing the bottleneck costs for the 2020 scenarios. The reduction due to wind is 28.6% for "2020 onshore" and 25.4% for "2020 offshore". Likely, offshore wind combined with a dedicated transnational offshore grid would give much lower bottleneck costs.

Table VI gives the results of the simulations for assessing the impact of wind variations on the bottleneck costs. In this case the simulations with wind in the

Table V: Bottleneck Costs Due Towind Power			
Scenario	**2005**	**2020 Onshore**	**2020 Offshore**
Simulations with wind power:			
Grid model cost [€/MWh]	25.5	34.4	33.8
Copperplate model cost [€/MWh]	23.5	33.9	33.3
Bottleneck cost [€/MWh]	2.00	0.45	0.47
Simulations without wind power:			
Grid model cost [€/MWh]	26.6	38.5	38.5
Copperplate model cost [€/MWh]	24.6	37.9	37.9
Bottleneck cost [€/MWh]	1.98	0.63	0.63
Δ bottleneck cost [€/MWh]	0.02	-0.18	-0.16
Δ bottleneck cost [€/MWh wind]	0.80	-2.44	-1.55

system are compared with simulations assuming constant generation from the wind farms. The bottleneck cost of these two assumptions (wind variations versus constant generation) appears quite similar, though a marginal cost (0.5 to 9.5% of the bottleneck cost) can be attributed to wind variations. Indeed, in this analysis startstop costs of thermal power plants are not explicitly included. Neither are costs related to balance management due to (wind) forecast errors.

Table VI: Bottleneck Costs Due to Wind Power Variations			
Scenario	**2005**	**2020 Onshore**	**2020 Offshore**
Simulations with wind power:			
Grid model cost [€/MWh]	25.5	34.4	33.8
Copperplate model cost [€/MWh]	23.5	33.9	33.3
Bottleneck cost [€/MWh]	2.06	0.45	0.47
Simulations with constant power:			
Grid model cost [€/MWh]	25.5	34.4	33.56
Copperplate model cost [€/MWh]	23.5	33.9	33.14
Bottleneck cost [€/MWh]	2.05	0.45	0.42
Δ bottleneck cost [€/MWh]	0.01	0.01	0.04
Δ bottleneck cost [€/MWh wind]	0.40	0.13	0.39

The flow within countries is assumed not constrained in this analysis. This may be right for some countries, but probably not for all. In particular the big flow within

Germany is noted, see Figure 8. This requires a transfer capacity within Germany of a magnitude that likely today does not exist. This may be taken as a weakness of the study, but has been unavoidable due to limited access to detailed data of country internal branch capacities. Results may thus be biased, but are still considered robust with regards to the relative impact of wind on transfer bottlenecks between countries.

Figure 8: Visualization of Average Power Flows in Northern Europe for the 2020 Scenario with 152 GW Onshore Wind Power and 48 GW Offshore Wind Power (2020 Offshore)

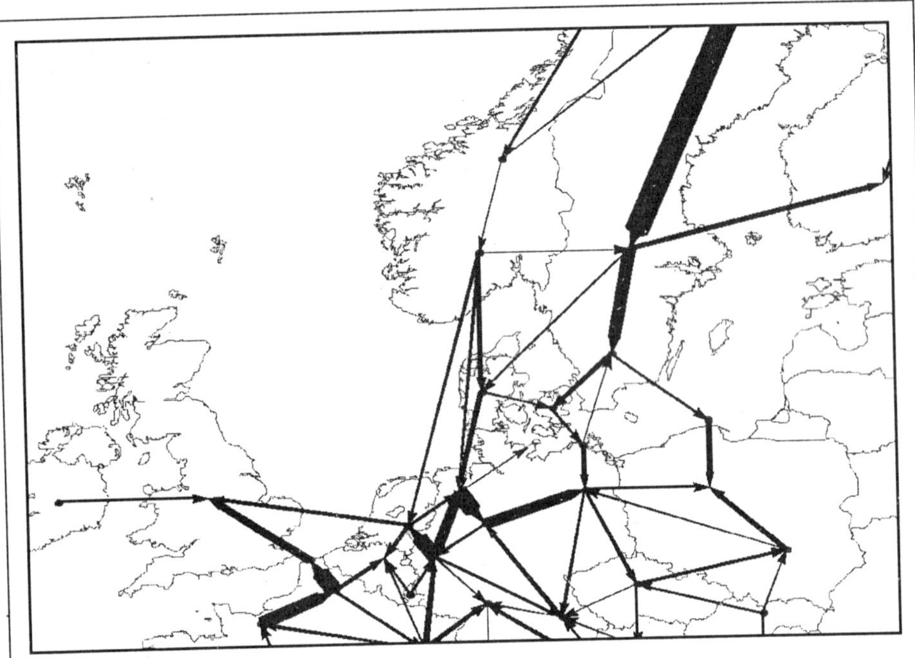

The Line Thickness Indicates the Magnitude of the Flow.

VI. Conclusions

Simulations have been carried out for studying the effect of offshore wind on power flows in the European high voltage network. The simulations have been prepared applying a detailed model of the European power system for year 2005 and for 2020 with 152 GW onshore wind capacity and 48 GW offshore wind capacity. The analysis shows that bottlenecks between countries exist, but are generally reduced from 2005 to 2020. Both the assumed grid development and wind power contribute to reducing the bottleneck costs for the 2020 scenarios. The reduction due to wind is 28.6% for "2020 onshore" and 25.4% for "2020 offshore" (relative to the situation in 2020 without wind). Likely, offshore wind combined with a dedicated transnational offshore grid would give much lower bottleneck costs. A modest cost (0.5 to 9.5% of the bottleneck cost) can be attributed to wind variations. Actual bottleneck costs experienced by system operators are expected mainly due to market constraints for trade between countries, and due to country internal limitations in transfer capacity.

Acknowledgment

The authors wish to acknowledge the European Commission (IEAA Programme) for financial support to the TradeWind project, and Cornell University for developing the software package MATPOWER.

(J O G Tande, L Warland, M Korpås and K Uhlen are with SINTEF Energy Research, Norway. The authors can be reached at: John.O.Tande@sintef.no, Leif.Warland@sintef.no, Magnus.Korpas@sintef.no and Kjetil.Uhlen@sintef.no respectively

F Van Hulle is with EWEA, Belgium. The author can be reached at frans.vanhulle @ewea.org).

References

[1] TradeWind, Further Developing Europe's Power Market for Large scale Integration of Wind Power, *www.tradewind.*

[2] M. Korpås, L. Warland, J. O. G. Tande, K. Uhlen, K. Purchala, and, S.Wagemans, "Grid modelling and power system data", TradeWind report D3.2, 2007, *www.tradewind.eu*

[3] Q. Zhou, J. W. Bialek. Approximate Model of European Interconnected System as a Benchmark System to Study Effects of CrossBorder Trades, IEEE Trans. Power Systems; 20(2), May 2005.

[4] G. Van der Toorn, "Wind power capacity data collection", TradeWind report D2.1, 2007, *www.tradewind.eu*

[5] J. McLean, "Characteristic wind speed time series", Tradewind Report D 2.3, 2007, *www.tradewind.eu*

[6] J. McLean, "Equivalent wind power curves", Tradewind Report D 2.4, 2007, *www.tradewind.eu*

[7] Pure power. EWEA. 2008, *www.ewea.org*

[8] F. Van Hulle, J. Dedecker, A. Woyte, "Mapping Offshore wind power capacity for TradeWind Grid Scenarios", 2007, *www.tradewind.eu*

[9] UCTE, "System Adequacy Forecast 2007-2020", Scenario B, January 2007.

[10] Union of the Electricity Industry. Statistics and prospects for the European electricity sector (EURPROG), 2005 and 2007.

[11] *www.nordel.org*

[12] *www.nationalgrid.com*

[13] P. Capros, L. Mantzos, V. Papandreou, N. Tasios. European Energy and Transport, Trends to 2030 – Update 2007. ICCSNTUA for European Commission, DG TREN, April 8, 2008.

[14] Norwegian Energy and Water Directorate, Quarterly report 4 -2005 (in Norwegian), p. 67, ISBN: 8241005768, January 2006.

[15] IEA. Interactive Recalculator, Downloads. *http://www.recabs.org/*, accessed 18/04/2008.

[16] Vattenfall, Annual Report 2006.

[17] Johan Löfberg, "mexclp.cpp", *http://control.ee.ethz.ch/~joloef/clp.php*

[18] *https://projects.coinor.org/Clp*

Section III

Experiences and Initiatives

12

Support for Wind Resource Assessment and Wind Farm Development in China

W L Wallace, Z Y Wang and S J Liu

China has an enormous potential for wind power development estimated at as much as 250 GW inland and another 750 GW offshore. Installed wind capacity in China in 2005 reached 1.26 GW with the addition of 498 MW, a 252% increase in new installations over 2004. Interest in the commercial development of wind power is also increasing with passage of the Renewable Energy Law in February 2005 and the establishment of strategic planning targets by the National Development and Reform Commission (NDRC) of 5,000 MW in 2010 and 30,000 MW in 2020 for wind development. Since the year 2000, the United Nations Development Program and the Global Environment Facility have cooperated with the NDRC to pursue a site characterization program to expand the wind resource information base for commercial development and increase the capacity in China for wind resource assessment and analysis. This paper will review support for resource assessment and trends for wind energy development in

China, and update information on the National Wind Concession Program, national planning targets, and policy support for commercial development from the Renewable Energy Law.

1. Introduction

Worldwide wind development has been accelerating with an average annual growth rate of 20% over the last five years, due to the favorable implementation of government policies in Europe and the US and the increasing recognition of the investment potential for wind by corporate, institutional, and private investors. Installed cumulative global wind capacity grew to 60,000 MW at the end of 2005, an increase of 11,407 MW from 2004, representing a 40% growth in new installations. While most of this growth took place in the US and Europe, strong development also occurred in Asia with India installing 1,253 MW to become the fourth largest market and China installing 498 MW to become the sixth largest market in 2005 [1].

1.1 Potential and Current Scale of Development of Wind in China

China has an enormous potential for wind power development estimated at as much as 250 GW inland and another 750 GW offshore for a total of more than 1000 GW. Cumulative installed wind capacity in China in 2005 reached 1,264 MW with the addition of 498 MW, a 252% increase in new installations over 2004, and China is now ranked eighth in the world in terms of cumulative capacity. This development has been accelerating in recent years, for example, during the last seven years the average cumulative installed capacity has been increasing at an average rate of 25% per year. During this period the development of wind power in China has made a transition from a heavy reliance on bilateral grants and soft loans to commercial development supported by domestic and international sources of investment.

The Chinese Government has also been providing support for an increasing scale of wind development through a wind concession program initiated in 2002. In the wind concession program, contracts are awarded on the basis of competitive bidding to establish the lowest price offer, after which power purchase agreements guarantee the electricity tariff for 15 years (30,000 full load hours), combined with other concessions in the form of providing site access roads, transmission lines to substations, and others. By the end of October 2006 there were nine concession projects awarded totalling 2,350 MW with tariffs of up to 6.4 US cents per kWh [2]. Through the concession program that require domestic content for wind turbines of at least 70% and other incentives, the Chinese Government has encouraged the development of a domestic base of manufacturers for wind turbines and components. China now has four domestic turbine manufacturers, of which Goldwind in Xinjiang province is the largest, as well as manufacturing plants opened by international companies such as General Electric, Nordex, Gamesa, and Acciona [3]. Wind turbines installed in new wind farms in China now range from 650 kW to 1.5 MW.

1.2. Future Prospects and Challenges for Wind Development

Interest in the commercial development of wind power in China has increased with passage of the Renewable Energy Law in February 2005 and with publication of implementing regulations for the law in early 2006. The National Development and Reform Commission will continue to support the national wind concession program, and electricity tariffs for large projects above 50 MW will continue to be set by competitve bidding procedures, resulting in power purchase agreements with guaranteed prices. Most new wind concession projects are now in the range of 100-300 MW. Local governments at the provincial level will also have flexibility for approving smaller projects (less than 50 MW) and can establish tariffs directly for projects having one developer or through competitive bidding for projects with interest from multiple developers. Provincial tariffs will be referenced to other wind farm projects within a given region. Future large scale concession projects will also link project development with the manufacturing of advanced wind turbine generators to support domestic manufacturing. In addition the Renewable Energy Law encourages capacity building in several areas, including initiating a national wind resource assessment program, providing support for domestic wind turbine manufacturing, establishing testing and certificaiton for turbines and wind farms, and exploring the potential for off-shore wind development. The National Development and Reform Commission has also established aggressive targets in its strategic national wind development plans, including increasing the cumulative installed capacity of wind power in China to 5,000 MW in 2010 and to 30,000 MW in 2020. In order to support these targets, the NDRC is also increasing China's capacity for regional planning for large-scale wind development in cooperation with the utility sector for aggregation of 1000 to 3000 MW of additional wind capacity in selected provinces. The NDRC is also currently preparing a National Wind Industry Development Roadmap in further support of these targets.

The near-term and longer-term targets for 2010 and 2020, while aggressive, are achievable given the wind potential in China and the existence of a rapidly evolving infrastructure for wind resource assessment, manufacturing, project development and project construction, wind farm management, and investment. This existing infrastructure will require ongoing improvements and will need to be expanded to meet the demands of the increasing development scales needed to achieve targets. One key area that represents a barrier to accelerated wind development is China's current capacity for wind resource assessment and site characterization, for which improvements are needed for wind measurement and data acquisition, wind flow analysis and modeling, GIS mapping, offshore wind measurements, and feasibility study preparation.

2. Support for Resource Assessment and Site Characterization

Since the year 2000, the United Nations Development Program and the Global Environment Facility have cooperated with the National Development and Reform Commission in the Project "Capacity Building for Rapid Commercialization of Renewable Energy in China" to assist in meeting China's national objectives for wind development. One of the Project's major contributions has been focused on capacity building to overcome one barrier to project development associated with the lack of high quality wind data. In 2002 the Project initiated a resource assessment and site characterization program to expand the wind resource information base for commercial development of wind in China.

Objectives for the wind support activities include: i) increase the general capacity of Chinese organizations to perform wind resource assessment and wind energy measurement projects at international best practice levels; ii) establish field installation and data acquisition protocols for the measurement of wind data, iii) develop advanced modelling and analysis capabilities for wind site characterization and feasibility study preparation, and iv) assist in developing technically and financially viable wind farm projects to promote wind commercialization in China.

The major components of the resource assessment and site characterization project include:

i) site selection, initially identifying 10 sites located in 8 provinces in northern, eastern and southern China,

ii) procurement and installation of equipment for wind measurements, including training for the installation and use of equipment,

iii) data collection and field monitoring by field organizations at each project site,

iv) data validation and wind flow analysis, and

v) preparation of site characterization reports.

2.1. Site Selection

During late 2001 and early 2002, the State Power Corporation (now reorganized into the National Grid Corporation) and the Long Yuan Electric Power Group Corporation investigated and identified ten wind farm development sites for inclusion in the project. Major criteria for site selection included potential for at least 100 MW of commercial development at each site, good potential wind resource based on nearby weather station data (if available), access to site, proximity to a local transmission grid, and others. For each site a field organization was identified with a potential commercial interest in the site to be responsible for installing equipment and collecting data, representing a source of cost sharing in the project. For each site,

Table 1: Identification of Sites and Field Organizations

Number	Site Name	Province	Field Organization
1	Yumen	Gansu	Gansu Jieyuan Company
2	Helan Mountain	Nigxia	Ningxia Electric Power Group
3	Huitengxile	Inner Mongolia	Inner Mongolia Wind Power Company
4	Dali	Inner Mongolia	Inner Mongolia Wind Power Company
5	Taonan	Jilin	Jilin Wind Power Company
6	Lichuan	Hubei	Lichuan Electric Grid Company
7	Poyang Lake	Jiangxi	Jiangxi Electric Power Company
8	Gulei	Fujian	Fujian Electric Design Institute
9	Putian	Fujian	Fujian Electric Design Institute
10	Xuwen	Guangdong	Guangdong Electric Design Institute

the preparation of project feasibility studies and commercial project development are the sole responsibility of the Chinese Government.

During 2002 the National Development and Reform Commission initiated the national wind concession program [2], and subsequently sites were candidates for inclusion in the national competitve bidding program. To date, one site, Huitengxile, has been developed based on inclusion in the national wind concession program.

2.2. Equipment Procurement and Installation

During 2002, the Long Yuan Electric Power Group Corporation was awarded a contract to manage the installation and data acquisition functions of the project and maintain quality control for data collection. Long Yuan was generally responsible for data retrieval, transfer, and archiving in a central database for the project. During 2002 until February 2003, the installation of 40 meteorological towers was carried out at the 10 project sites, under subcontracts between the Long Yuan Corporation with the eight field organizations to carry out the installation and data collection activities. The field component of the wind resource assessment program involved training of field organizations, equipment procurement and transport, installation of systems, development of data collection and data handling protocols, and data collection and transfer to a central analysis group. A quality control program provides certification of the work of the field centers according to project standards established during a resource assessment and training workshop conducted at Huitengxile in August 2002.

Global Energy Concepts (GEC) from the United States was the lead consulting group that designed the tower configurations and did the training for the Long Yuan Corporation and the local field organizations in Huitengxile, Inner Mongolia. The National Renewable Energy Laboratory (NREL) of the United States also provided assistance for the design of the guidelines used for wind resource assessment in China.

2.3 Field Configurations

A total of 38 measurement towers have been installed at the ten project locations, with the relocation of two 40-meter towers in 2003 to two additional sites in Guangdong province to collect wind data for sites targeted for the NDRC wind concession project. A typical installation configuration consists of one 70-meter tower

(Rohn 45 GSR series lattice towers) and three 40-meter towers (NRG 40 meter Tall Tower masts) with redundant anemometer and wind vane sensors. The measurement program is designed to be flexible, since the terrain and topographical configurations differ considerably among the ten sites. The 70-meter towers are fixed and are designed to collect long term data for future reference purposes. The tilt-up 40-meter towers are movable and will be relocated to other locations by the field organizations in charge of the equipment at the end of the measurement program. Table 2 provides the standard configurations of the 70-meter and 40-meter towers used in the project.

The measurement period was conducted for all sites during a two-year (24 month) period within the timeframe of mid-2002 to March 2005. During the measurement period, periodic reviews of the tower installations were performed with local consultants, the Long Yuan Electric Power Group Corporation, and the NDRC/UNDP/GEF Project Management Office. This enabled the project to have a continuing monitoring capability for the equipment operation and local counterpart on-site training to make sure appropriate protocols are being followed. As shown in Table 2, considerable redundancy was built into the tower sensor configuarations, since field visits needed to be conducted approximately every two months to collect data, due to the limited capacity of the NRG 9300 dataloggers used for the project. If the

Table 2: Configuration of the 70-Meter and 40-Meter Measurement Towers

	Configuration of 70-meter lattice tower		Configuration of 40-meter tilt-up tower	
Measurement	Data Logger A	Data Logger B*	All inland sites	Guangdong and Fujian
Wind Speed NRG #40C Anemometers	70 meter	70 meter	(2) 39 meter	(2) 39 meter
	60 meter	60 meter	(2) 25 meter	(2) 25 meter
	50 meter	50 meter	(1) 10 meter	(1) 10 meter
	40 meter	40 meter		
	25 meter	25 meter		
	10 meter	10 meter		
Wind Direction NRG #200P Windvanes	70 meter	70 meter	39 meter	39 meter
	40 meter	40 meter	25 meter	25 meter
			10 meter	10 meter
Temperature	3 meter	3 meter	3 meter	none
Barometric Pressure	4 meter	4 meter	none	none
*Redundant				

project were initiated today, less expensive dataloggers with greater data capacity are available to improve the reliability of data collection. It is also feasible to use mobile phone modem connections to download data from most regions in China today.

2.4. Data Analysis

Validation and analysis of the wind data from the ten project field sites was performed by the China Water Resources and Hydropower Consulting Corporation in Beijing. International consultants were also contracted to assist the Hydropower Corporation. Risoe National Laboratory in Denmark provided assistance in applying the WAsP software for wind flow analysis and modelling to the wind data and Garrad Hassan provided a review of the results of site characterization analysis for quality control and certification of the results.

A one-year validated database was generated for each site using the most complete contiguous 12-month data sets for wind speed and direction within the raw two-year database. Special software was developed by the Hydropower Corporation and reviewed by Risoe to perform the data validation using criteria recommended by GEC, Risoe, Garrad Hassan, and Hydropower consultants. In 2005 the Hydropower Corporation trained provincial research institutes on the use of this software and now offers the software for sale.

Basic site characterization was performed using the WAsP software to perform analyses on the validated data generally using the datasets from the highest sensors, except for wind shear analysis. Analyses included determining the average annual wind speeds, wind speed and wind power density monthly (seasonal) and diurnal variations, calculating wind direction Rose curves, and wind speed and energy distributions. WAsP wind flow modeling validation analyses were performed to cross correlate predicted wind speed between measurement masts and to determine predicted error ranges. Where weather station data was available nearby some sites, correlations were also performed on 10-meter mast data available from national weather stations to attempt to extend the historical wind performance regime for sites to ten years or more. Wind shear analyses were also performed. The data generated would be used in the course of performing feasiblity studies for project development, but the preparation of feasibility studies fell outside the scope of the project.

2.5. Site Development

Some results from the wind modeling and analyses for site characterizations are shown in Table 3, as well as a summary of some of the developments which has occurred at sites during the last few years of the project.

Average annual wind speeds are determined at 70-meter mast heights, except where noted, and wind class is based on the China national standard GB/T 17910-2002.

By the end of 2005, 283 MW of wind power has been developed at four of the project's sites. Most project development has been promoted at the provincial level, except for Huitengxile, which was included in the NDRC's 2002 national wind concession program. Another 430 MW is in various stages of planning pipelines at the provincial level for 100 MW at Yumen, 50 MW each at Taonan, Putian, Gulei, Poyang, and Xuwen, and 80 MW at Lichuan.

Table 3: Wind Site Characteristics and Development

Site	Province	Avg. Ann. Wind Speed (m/s)*	Wind Class (GB/T 17910-2002)	End of 2005 (MW)	Developer
Huitengxile	IMAR	8.2	4	68.5	Beijing International Electric New Energy Co., Huadian IMAR Wind Power Co.
Yumen	Gansu	7.7	4	52.2	Gansu Jieyuan Company
Helan Shan	Ningxia	6.6	3	112.2	Ningxia Electric Power Group
Taonan	Jilin	7.0	3	50	Datang Jilin Electric Power Generation Co.
Putian	Fujian	8.4	3	NA	NA
Gulei	Fujian	6.3	3	NA	NA
Poyang Lake	Jiangxi	6.0	2	NA	NA
Xuwen	Guangdong	5.7**	2	NA	NA
Lichuan	Hubei	5.5***	2	NA	NA
Dali	IMAR	6.5	3	NA	NA

* 70-Meter Height; ** 60-Meter Height; *** 39-Meter Height, NA-not applicable

Conclusions

As wind energy development continues in China, increased scrutiny will be made on any new projects to determine the quality and reliability of the wind data and the data analysis work. Previous wind farm analysis work done in China has created expectations for energy production at some sites, which have not been met after construction of the projects. This circumstance has caused some projects to be uneconomical. This problem is avoidable with proper data collection and analysis.

The strengthening of the field procedures and site characterization will give developers and financial institutions more reliable information and lower perceived risk in the financing of these projects. As the industry continues to develop, more project developers will be able to meet these requirements and find a more receptive investment environment.

Strengthening capacity in China for wind resource assessment and analysis is also essential if China is to meet its aggressive targets for wind development by 2010 and 2020. In 2005 the NDRC has established a national wind resource assessment program in oder to support capacity building and update the national wind resource mapping and estimates for wind power potential in China.

(William Wallace, Senior Project Leader, is a member of the Market and Policy Impact Analysis Group in the Strategic Energy Analysis Center. Author can be reached at william_wallace@nrel.gov

Z Y Wang, and S J Liu ;NDRC/UNDP/GEF Project Management Office).

References

[1]. A. Cameron, *Renewable Energy World*, Vol. 9, No. 4, 2006, pp. 56-66.

[2]. J. Ku, D. Lew, P. Shi, and W. Wallace, *Renewable Energy World*, Vol. 8, No. 4, 2005, pp. 212-223.

[3]. E. Martinot, *Renewable Energy World*, Vol. 9, No. 4, 2006, pp. 45.

13

Institutional Factors and Knowledge Co-Evolution for Developing the Wind Energy Industry in India

Rekha Rao

In this article we explore the factors that promote the development of nascent wind energy sector in India leading it to become one of the important wind energy producers in the world. We investigate this with respect to India's unique policy of joining hands with other countries with advanced wind energy sector to develop its own wind energy industry. In this study we concentrate on the relationship between India and Denmark. We highlight the institutional factors that were responsible for development and sustenance of the wind industry in India. Our analysis leads us to believe that knowledge co-evolution within and between institutions related to wind industry was responsible in helping India to establish itself as the fifth largest wind energy producer in the world.

1. Introduction

Since the early 1980s, some manufacturing of wind turbine was undertaken in India, but the actual impetus was provided by the initiatives undertaken by the Indian government for development of wind energy sector in 1985. Also, the policy reforms in the 1990s allowed foreign firms to increase their equity share above 40 per cent and in some sectors there are even 100 per cent-owned multinational subsidiaries. One of the results of this liberalization policy has been the substantial presence of foreign firms in the Indian wind turbine industry. Since then the wind energy sector has undergone one turbulent period after another, yet the Indian wind energy sector has managed to grow into the 5th largest in the world (GWEC, 2005) and a leader in Asia. According to the Wind Energy Division of the National Aerospace Laboratories (NAL), nearly 40 percent of India's installed electricity capacity of over 120,000 MW could be generated from wind energy. Today India only gets 2.7% of its power from wind energy[1]. Thus, there still exists a tremendous opportunity for growth of wind turbine industry in India.

At the same time, owing to the technological advancement, the cost of wind energy has fallen tremendously. Yet, the cost of wind energy is higher than the conventional energy (Wong, 2005). Due to this reason government has been proactive in developing the wind energy industry via means of government supported R&D, subsidies and incubation of new start-ups in this sector. Nevertheless, due to high degree of uncertainty, implementation of a successful technology policy[2] in an emerging industry is problematic. The benchmarking or copying of already successful technology policies is not likely to reap substantial benefits (Maskell, 1996, Lundvall and Tomlinson, 2000, 2002). Although the imitations of the policies are successful, the outcome will differ because the policies and institutions interact with firms and are rooted in distinct national and industrial settings. Given these difficulties, it is remarkable that India has in a relatively few years managed to position itself as one of the leading nations in wind energy.

We highlight the institutional factors that were responsible for development and sustenance of the wind industry in India. Also, in this article we posit that India's policy of joining hands with other countries with advanced wind energy sector to

1 *http://www.deccanherald.com/deccanherald/oct42005/snt.asp* (as accessed on October 4th, 2005).

2 Technology policy refers to policies that focus on technologies and sectors. Technology policy often focuses on rapidly growing markets that are characterized by high rate of innovation (Lundvall and Borras, 2005).

develop its own wind energy industry has helped it tremendously. Primarily, we concentrate on the relationship between India and Denmark. This is so because Denmark is the leading wind energy producer in the world and has consistently lead innovation in wind energy industry (refer to K ristinsson and Rao, 2007, 2008 for more details). Our analysis leads us to believe that 'knowledge co-evolution' within and between institutions related to wind industry was responsible in helping India to establish itself as the fifth largest wind energy producer in the world. The knowledge co-evolution between a developed and a developing country is different from basic 'technology transfer' between them. Bell (1984) describing technology transfer between transferor and recipient talks about the knowledge involved in this process. Author suggests 4 modes of technological knowledge transfer between these entities, one, it can be in form of technological artefacts or accompanying manuals; second, it can involve reverse engineering on part of the recipient; third, the transfer takes place in form of formal training sessions by suppliers to recipient and finally, it can involve 'learning-by-doing' on part of the recipient company. In this process, the recipient gains during the transaction in terms of knowledge accumulation, whereas, the transferor can be assumed to be a passive knowledge disseminator and observes no changes in his core knowledge capital. On the other hand, Lundvall (1985, 1988) describes 'interactive learning' as a process in which those involved increase their competence while engaging in the innovative process. This relationship can also be described as a mentor/apprentice relationship in which both parties benefit from increased interaction. In our study, we suggest that this co-evolution of knowledge has lead to rapid development of wind industry in India.

Given the difficulties in collecting primary data on the Indian wind energy industry, and especially on knowledge co-evolution between Indian and Danish actors, the study's contribution is based on secondary evidence. Heaton (1998) mentions the limited opportunities for conducting primary research and also suggests that the costs of qualitative work should promote researchers to consider maximizing the use of secondary data available to them. Several other authorships (Hinds, Vogel and Clarke-Steffen, 1997, Sandelowski, 1997, Szabo and Strang, 1997, Thorne, 1990) have suggested positively the use of secondary analysis of qualitative studies. The main question which arises is whether the data that we have been able to gather is amenable to secondary analysis, and we can confidently say that in our case it does.

In the next section, we will discuss the institutional factors that were conducive for development of wind energy sector in India. We will look at the interaction between Danish and Indian actors in this sector. In section 3, we will look at the know ledge co-evolution between and within the Danish and Indian wind energy sector. Section 4 concludes.

2. Institutions and Technology Policy

Government-lead Development of Wind Energy in India

Compared to Denmark, India was a late convert to wind energy development. Yet, by mid-1990s there was convergence between the wind energy production in Denmark and India. Figure 1 below shows the development of wind power in Denmark and India. We see that the initial wind power production in India was mainly demonstrative in nature, where the government was the major instrument. Thus, one of the major players in the development of the wind energy sector in India has been the government and its wind energy friendly policies.

Figure 1: Wind Power Development in Denmark and India

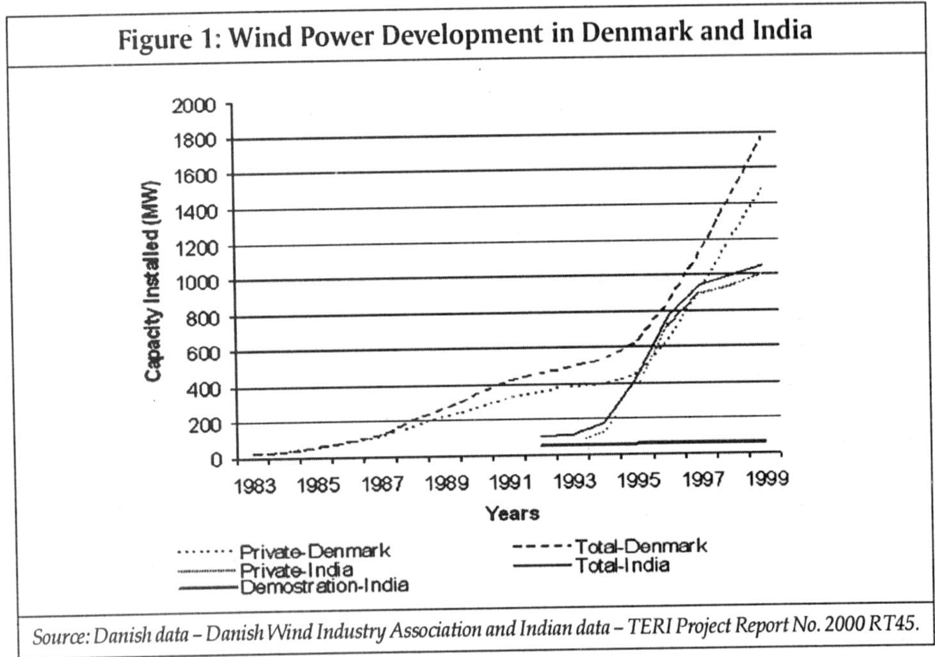

Source: Danish data – Danish Wind Industry Association and Indian data – TERI Project Report No. 2000 RT45.

The position of the Indian government as a facilitator for private interest and development of the Indian wind turbine industry can, to some extent be attributed to its relations with the Danish government. This interaction resulted in the development of several projects for promotion of wind power sector in India. The Centre for Wind Energy Technology (C-WET) based in Tamil Nadu and established in cooperation with the Danish government, has acted as a technical focal point for wind power development in India. Indeed, the biggest Indian state in wind energy is by far Tamil Nadu [3], in the south of India. In 1986, the first wind farm with ten 55 kW wind electricity generators (WEGs) was started in Tamil Nadu. The state has also received major funding assistance from Ministry of Non-conventional Energy Sources (MNES). In addition, the local cost is shared by Tamil Nadu Energy Development Agency (TEDA) and Tamil Nadu Electricity Board (TNEB) equally. This has lead to Tamil Nadu making a rapid progress in development of wind energy in that state. It has so far established demonstration wind fams at eight areas with a total capacity of 19.10 MW. TEDA has undertaken state-wide wind resource assessment programme right from 1986 with the funding assistance of Government of India (MNES) and government of Tamil Nadu and with the Technical Assistance of Indian Institute of Tropical Meteorology, Bangalore. Moreover, the programme of installation of windmills for water pumping is being carried out in Tamil Nadu with the subsidy from the funding assistance of MNES and government of Tamil Nadu. These various initiatives have been the major sources of the tremendous growth Tamil Nadu has made in the development of wind power.

Some other parallel efforts were made by other governmental organizations along with the above incentives of government. For instance, the National Aerospace Laboratories (NAL) has been involved in the development of wind turbine development programmes which being partly funded by Danish actors through the Centre for Wind Energy Technology (C-WET). There are difference between the wind speeds in India and other parts of the world. In India, wind speeds are much lower than those wind speeds in Europe and lower speeds drastically affect the wind power output. Thus, a turbine has to be properly rated to efficiently generate electricity at these lower speeds, otherwise, it is not going to work well in India. Hence, NAL is involved in building wind turbines that are adapted to the conditions in India[4]. Applying wind turbine technology in different context has therefore generated new knowledge to the benefit of all involved.

3 Tamil Nadu numbers from the Indian wind energy association webpage.

4 *http://www.deccanherald.com/deccanherald/oct42005/snt.asp* (as accessed on October 4th, 2005)

Stimulated by the success of the above several government programmes, private firms has made rapid strides on wind power generation. The first private sector wind farm of the country was set up in Tamil Nadu during 1990 with two wind electricity generators of 250 KW each at Muppandal. The private sector has from these modest beginnings grown tremendously (Figure 1).

The Role of the Utilities

Both the Indian and Danish governments have had to exert their authority to assist the development of wind turbine industry. In spite of the fact that the markets for electricity in Denmark and India are dissimilar in many aspects, both countries have, arguably successfully, taken a path of government involvement for development of this nascent industry.

In India, one of the challenges to development of the wind energy sector is the considerable resistance from the State-run utilities to grant "third party sale" [5] facility to wind power producers (Rajsekhar *et al.*, 1999). The utilities see this as competition for their most high end industrial customers. Complicating the situation is the presence of the State-run utilities as the primary customer of wind energy producers. On the other hand, the utilities see the wind energy producers as a peripheral supply option and possibly as a nuisance because of their low reliability (ibid.). The Indian Government has however in recent years worked to solve these complications for the benefit of the wind energy industry.

University Relations

As India was not at the same level as Denmark in research in wind energy, a part of the Danish competence in wind energy has been transferred to India through educational facilities such as C- WET. Due to this India has managed to build up significant competences in wind energy technology and wind project execution. Denmark was the first country to promote aggressive quality certification and standardization in wind turbine knowledge (Lewis and Wiser, 2005). In the late 1970s the Risø National Laboratory began approving turbine design to ensure reliability and safety standards were adhered to (Sawin, 2001). This process was implemented with the help of owners, manufacturers and the Danish authorities

5 Third party sale amounts to allowing wind power producers to use the grid infrastructure to sell power to any industrial client at any mutually agreed rate.

who all had an interest in ensuring the quality of Danish wind turbines. This attention on quality has also been adopted by Indian wind energy players and as a result India is now one of the foremost countries in wind turbine quality control. Local educational institutions have also played an important role in building up the right competences for the development of the wind energy industry. In India, there exists a high degree of interaction between the government agencies and the universities in wind turbine technology. One of the more flourishing ones has been the case of TEDA and Indian Institute of Tropical Meteorology, Bangalore. State-wide wind resource assessment programme has been undertaken by TEDA right from 1986 with the funding assistance of MNES and government of Tamil Nadu and with the Technical Assistance of Indian Institute of Tropical Meteorology, Bangalore.

Wind Energy Policies in India

Generally speaking there are two views on technology policy. The first view, which can be termed "pro-market view" claims that profit maximizing firms driven by competitive pressure, will develop technologies that are both profitable and beneficial to society. The proponents of this view argue for a market based technology policy with the possible exclusion of subsidization of R&D. Technological change is seen as exogenous to the system. Hence, this view of technological change is really not a view on technological development and innovation but about the efficient utilization of technologies, given the state of technological knowledge. The other view is the "state promotion view" that identifies the many market imperfections that slow down the pace of technological innovation. The supporters of this view argue that policy interventions are needed to generate incentives for growth and innovation, because the process of technological adaptation is subject to a number of important market imperfections. The wind turbine industry technology policy has been characterized by the state promotion view, as generation costs of wind energy has historically been considerably higher then that of conventional energy sources. Globally, the role of the state has therefore been prominent in getting the industry through its emerging phases.

During the 7th National Five Year Plan (NFYP), which ran from 198 5 to 1990, Indian Government gave incentives to grid quality power generation by wind turbine technology. Wind power generation became the thrust of India's Ministry of Non-conventional Energy Services (MNES) by the start of the 8th NFYP.

MNES has formulated a series of policy incentives and fiscal incentives that have been successful in the development of the wind power sector. Along with this policy, individual State Governments declared their own incentives. These actions by the State Governments created a conducive investment environment that led to a surge of activity in this sector. The fiscal incentives extended by the Indian Government to the wind turbine sector are of twofold. Direct taxes – 80 per cent depreciation in the first year of installation of a project, and a tax holiday for 10 years. One outcome of these incentives has been to encourage industrial companies and businesses to invest in Indian wind power. The main reason is that owning a wind turbine assures them of a power supply to their factory or business in a country where power cuts are common. Mainly because of this the private sector has dominated investment, to about 97 per cent (GWEC, 2005). Thus, in India, wind farms often consist of clusters of individually owned generators. Governmental initiative for providing financing has helped this sector to gain foothold. The Indian Renewable Energy Development Agency (IREDA) has played a significant role in the promotion of wind energy, attracting bilateral and multilateral financial assistance from world institutions and the private sector. There are but few indigenous private players in this sector which are actively pursing the process of innovation and adaptation of the foreign technology to meet the local needs and seek international accreditation. In the beginning, financing institutions other than IREDA were not willing to invest in wind power projects due to lack of exposure and experience in this sector. In 1993-94, the World Bank provided financial assistance of $43 million for wind energy to IREDA. This opened the window for large-scale financing and growth of the industry in the mid-1990s. The growing interest among wind project sponsors encouraged other financial institutions to begin financing wind projects, including the Industrial Development Bank of India. (IDBI), the Industrial Credit Investment Corporation of India (ICICI) and the Gujarat Industrial Investment Corporation Limited (GIIC).

3. Knowledge Co-Evolution between and within the Danish and Indian Wind Energy Sector

Actors and an institutional set-up are undoubtedly important for technology development, but without interactions nothing will happen (Lundvall, 1988). By establishing their subsidiaries in India, Danish firms were able to interact with the local Indian firms. They were also able to develop several joint ventures with

Indian companies. One such example of this type of relationship is Vestas RRB India Ltd. which was set up in 1987 as a joint venture between the Danish firm Vestas and RBB, an Indian firm. Danish firms also carried out several large investment projects in India such as Vestas Wind Systems's six DANIDA wind energy projects in India 1988. These activities of knowledge Generation and assimilation were also supported by the Government in both countries, thus leading to knowledge co-evolution. In 1987 the governments of Denmark and India established a collaboration programme that conceived a 20 MW wind energy demonstration project. The project included a 10 MW wind farm in Gujarat and two wind farms (4 MW and 6 MW) in Tamil Nadu. In 1990 the wind farms were successfully commissioned. The success of this project led to cooperation between Danish and Indian firms by establishing production facilities for wind turbine generators in India. As a result of this development the need for strengthening the associated technological infrastructure with respect to the quality and standards of wind turbines became clear.

Also, these collaborations created an environment for Danish firms to envision long term investments in Indian wind energy market. Since the mid-1980s many Danish firms have been present in India. Their presence in India is therefore neither short term nor an one off transaction. A testament to this attitude is how NEG Micon reacted when faced with a severe problem in India. Instead of cutting their losses NEG Micon decided to start a 100 per cent subsidiary in Chennai in 1997 when its partner, NEPC, could not service the 1,200 turbines that had already been installed in India. Also, the Danish and Indian governments have made sustained investments for development of this industry in India. Among them being The Centre for Wind Energy Technology (C-WET) based in Tamil Nadu. As elaborated on earlier C-WE T has acted as a technical focal point for wind power development in India.

As a part of this effort educational institutes were set up as a fruitful outcome of the interaction between India and Denmark. These institutes served to build thrust and a common language between actors in India and Denmark. In 1994 two Environmental Training Institutes (ETIs) were established under the Danish cooperation programme in the Karnataka and Tamil Nadu states of India. The purpose of the programme has been to equip the ETIs with training facilities and to develop local trainers who in turn provide training to regulatory authorities, municipalities or companies. The ETIs have developed 24 training courses on several environmental themes and have trained more than 5000 participants. ETI Karnataka has acquired ISO 9000 certification. ETI Tamil Nadu is also planning to do so. In 1997 the Government of India and the Government of Denmark agreed to establish a national wind turbine test station under Centre of Wind Energy Technology (C-WET). The aim of this project is to build and strengthen the Indian capacity for testing, certification and quality assurance aspects of wind turbines. As a result of this policy India is now considered one of the three world leaders in this area of the wind turbine industry[6].

As can be seen from these examples there was considerable elements of non-market institutional support, with a continued involvement of Indian and Danish Governments that served as intermediaries in development of wind energy sector in India. The presence of these non-profit oriented bodies like State Governments served as a signaling mechanism for building trust within private players in this sector. This was especially important for further development, because of the crucial role of trust for interactive learning. It also hard to imagine that the educational and other non-profit organizations could have been successfully established and run without a substantial amount of trust between the actors involved.

In both Denmark and India, initial knowledge and learning processes were generated in wind energy projects that were demonstrative in nature and/or developed by the Danish and Indian Governments in collaborative efforts. As the experience and knowledge on how to build such projects became established, several local and international players stepped in to build on the knowledge that was already generated (Figure 1). Therefore the transfer of knowledge was not a singular event but rather a cumulative process in which new knowledge was build on what was already established. Learning is partially cumulative. What one learns depends on what one already knows and in that sense the production structure affects the learning processes in the economy. Initially, the Indian players are likely to be behind the Danish firms on the technological frontier, but when one considers the fact that wind turbine industry is a low-tech sector, this might not seem like a major hindrance for development of local knowledge for wind turbine manufacturers in India. Indeed, we recently see that there are several local wind turbine players that are actively gaining foothold in the global market.

---

6 The others being Denmark and Germany.

Indeed, the knowledge that was generated in the Danish wind energy industry could not be used 'as it is' in the Indian market as there was, among other things, difference in the wind speeds in India and Denmark. Therefore a bidirectional process was needed to make the technology work in the Indian context. Thus, new technology had to be developed and present technology adapted, to meet the local market needs in India. Rosenberg (1982) points out that, particularly in the case of technologies that consist of several interacting parts and that have to function in changing environments, learning by using is essential. Wind turbines are a typical example of this kind of technology (Kamp, 2006). As mentioned before, Indian wind speeds are much lower than European wind speeds. Unless a turbine is properly rated to efficiently generate electricity at these lower speeds, it is not going to work well in India. Hence, Danish firms were obligated to learn and develop new technologies that were suitable for low wind speeds. Thus, both parties learned in the transaction. This was complementary to the new know ledge that was being assimilated by the Indian wind turbine producers in joint ventures like those between Vestas and RRB (India) Ltd. Complementary to the learning of both parties was the atmosphere of experimentation and openness that prevailed. An example of this is that private Danish players undertook development of wind turbines in an uncertain emerging market with new local players. In this process there was open technology based interaction between Danish and Indian private players during the installation and maintenance of the wind mills.

In India, Danish designed windmills have played an important part in the learning and knowledge generation within the industry. In 1986 wind farm activity started with a boom in India. Five wind farms at Mandvi, Okha, Devgarh, Puri and Tuticorn were installed. These wind farms used mostly the second generation of wind turbines. In 1988, in the second phase of the program, wind turbines from the third generation were introduced (Singh, 1998). These and other wind farm developments in India have substantially benefited from the high degree of Danish involvement and have created local knowledge in wind energy development. Thus, the user has gotten access to technical information and has build user competence. These benefits have come both through the use of Danish wind technology and through knowledge transfers in demonstration projects, additionally the capabilities gained through the build up of C-WET have been very important for the development of the Indian wind turbine industry . Learning by interacting, instead of technology transfer, has enabled firms to 'draw' knowledge from external sources and create their own competences. An example of this is the growth of indigenous Indian actors like Suzlon energy that has almost 6 per cent of the global market share. The competence developed in India is also confirmed by leading industry players such as Ramesh Kymal at NE G Micon (India). "Not only did the high cost involved in manufacturing from Denmark tilt the decision in our favour, the quality of the products that went from here too helped us. I would say the quality of the products that went from here made them think of starting a manufacturing base in Chennai for the world market, strongly…"[7].

In this section we have argued that the development of the Indian wind industry is characterized by knowledge co-evolution between the Indian and Danish wind sector players. In this development Danish actors have played a leading role by helping to build a long term partnership that has benefited both Danish and Indian actors. We however do not claim that building such relationship is an easy task. For the development of the Indian wind sector the Danish and Indian Governments played a vital part by taking the first steps and by introducing the relevant actors. Building on this work, private actors have managed to build a truly impressive wind industry in India.

Conclusion

In this article, we attributed current success of Indian wind energy development to the development of efficient institutions and implementation of effective policies and to the knowledge co-evolution in this sector due to India's outward looking initiatives. We mainly focused on India's extensive relationship with Denmark, which is a pioneer in wind energy technology. This paper highlights the role of technological policy and institutions in determining the success or failure of industrial development. We put special emphasis on development of suitable institutions and implementation of secondary process which see to the efficient working of these institutions. Despite having developed in very different national and cultural contexts, many important actors within the wind energy industries in Denmark and India have cooperated, to the benefit of both parties. The effect of this cooperation has been to develop a vibrant Indian wind turbine industry, but along with this it has without a doubt also benefited Danish firms in their efforts to establish themselves in a growing market outside Denmark.

---

7 *http://www.rediff.com/money/2003/jun/26neg.htm*

At the government level the early efforts to establish the Indian wind turbine demonstration projects had an important effect in attracting private investors to wind energy projects. The cooperation programme between the government of Denmark and the government of India to establish Indian wind farms and production facilities provided the basis for the growth of an emerging industry. The governmental intervention in terms of policies in the primary stages of the industry acted as a solid foundation for the industry to find base among complementary institutions like financial markets (for example, soft loans for developer of wind farms in India) for development of the sector.

At the firm level we see the strong presence of Danish firms in the Indian wind turbine industry. In 2003 Danish firms had a 38.1 per cent market share in India. LM Glasfiber, the leading Danish rotor blade supplier, has also established itself with its Asian base in India. The strong presence of Danish firms in the Indian wind turbine industry is now being matched by Indian manufacturers. Suzlon Energy, the largest Indian wind turbine manufacturer, has recently established its international headquarters in Denmark due to its base of wind energy expertise and extensive network of component suppliers (WPM, October 2004:25). Thus the indigenous Indian firms that were established due to the favourable government policies were not only able to meet the local demand, but have become successful players in international market. One major reason for this is that firms like Suzlon Energy have been able to meet the needs of the developed markets both in terms of quality and price. For example, Suzlon Energy have R&D centers in Europe (Netherlands and Germany) and hence have been able to be at the frontiers of technological advancement, plus they have development and production facilities in India and China and hence, has been competitive on price. Although at present it appears more of an export based industry, India does have a large, but mostly untapped domestic demand. Even though the wind speeds in India different from those in Denmark, it is hard to imagine that the nature of product developed for Indian market will be so localized that it would lead to segmented domestic and export markets. For example, out of 7 power classes for wind speeds in US we observe that many of these wind speeds overlap with the wind speeds in India[8].

[8] *http://www.awea.org/faq/usresource.html*

At the policy level we see similar emphasis being made in both Denmark and India. Examples of this include the high awareness of the importance of having stable policies, as well as an emphasis on importance of utilities and grid connections for the early development of the industry. Additionally, in Denmark a combination of early R&D efforts and stringent certification standards, were the primary policy drivers in developing a large wind turbine industry. This practice has been transferred to India through the establishment of C -WET. As a result India is now one of the three world leaders in certification and testing of wind turbines, despite being a developing country. Thus a nascent sector in developing economy experienced positive effect of interaction with the same sector in developed economy, both at sectoral as well as firm level.

We argue that knowledge co-evolution with the Danish players, for example, in form of partnership of Danish firms with the local Indian wind turbine manufacturers, Danish companies establishing their subsidiaries in India, development of technical institutes to facilitate knowledge and capability building in wind energy technology in India, has benefited both the Danish as well as the Indian firms. This knowledge co-evolution has helped Danish firms to develop wind turbines and other technologies to match the wind speed experienced in Asia which is much lower than that seen in Europe. Thus, the kind of knowledge generated (wind turbines needed to be developed that were suitable for Indian needs by local or foreign manufacturing firms) and mode of transfer of knowledge (the new wind turbines had to be tested by local engineering staff who had to follow local regulatory guidelines) suggested that it is better communicated in face-to-face interaction between people and that a social context is essential for constructive knowledge diffusion by way of learning-by-interaction (Lundvall and Johnson, 1994).

Indeed, it is important to not only focus on if certain actors interact, but also how and why actors interact. A natural extension of such a focus would be an investigation into the role of trust, fairness and culture for industrial development. From a learning and innovation perspective this could be a crucial explanatory factor.

Notwithstanding some limitations, one of which happens to be the lack of quantitative analysis, we feel that the strength of our approach has been to bridge the gap in knowledge on how the wind energy industry in India developed. We are able to map the various institutions and actors in wind turbine industry in India and Denmark. Our results are exploratory in nature and should be seen as initial research conducted to clarify and examine the problem (Schutt, 2006). That is, why was the Indian wind energy industry so successful? This research will then be used as the basis for delving further into the subject, through in-depth case studies on individual firms involved in this industry. Nevertheless, our study is a definitive step forward in understanding the learning process between Danish and Indian agents and the evolution of the Indian wind energy sector. We find that India's policy of attracting world class actors to advance its emerging wind energy industry has without a doubt contributed substantially to its successful development.

(Rekha Rao is Research Associate at Imperial College, London. The author can be reached at rekha.rao@imperial.ac.uk).

References

Bell, M. (1984) "Learning and the accumulation of industrial technological capability in developing countries". In: Fransman, M., King, K. (Eds.), *Technological Capability in the Third World. Macmillan,* London, pp. 187–209.

GWEC, Global Wind Energy Council (2005), Wind Force 12, *http://www.gwec.net/fileadmin/documents/Publications/wf12-2005.pdf*

Heaton, J (1998) Secondary Analysis of Qualitative Data. UniS—Social Research Update, 22. Available at: *http://sru.soc.surrey.ac.uk/SRU22.html* [Date of Access: 10.12.2007]

Hinds, P S, Vogel, R J, Clarke-Steffen, L (1997) "The possibilities and pitfalls of doing a secondary analysis of a qualitative data set", *Qualitative Health Research,* 7(3), 408-24.

Kamp, L M (2006) "Danish and Dutch wind energy policy 1970-2000: lessons for the future", *International Journal of the Environment and Sustainable Development,* 5 (2), 213-220.

Kristinsson, K and Rao, R (2007) "Learning to Grow: A Comparative Analysis of the Wind Energy Sector in Denmark and India", DRUID Working Paper no. 07-18.

Kristinsson, K and Rao, R (2008) Interactive Learning or Technology Transfer as a Way to Catch-Up? Analysing the Wind Energy Industry in Denmark and India, *Industry & Innovation,* vol. 15(3), pages 297-320.

Lewis, J and Wiser, R (2005) Fostering a Renewable Energy Technology Industry: An International Comparison of Wind Industry Policy Support Mechanisms,

Environmental Energy Technologies Division, Ernest Orlando Lawrence Berkeley National Laboratory, Download from *http://eetd.lbl.gov/EA/EMP*

Lundvall. B.-A. (1988) 'Innovation as an interactive process – from user-producer interaction to the national system of innovation'. In: G.Dosi *et al.*, (Eds), *Technical Change and Economic Theory* . London: Pinter Publishing

Lundvall, B.-A. (1985) *Product innovation and user- producer interaction.* Aalborg: Aalborg University Press.

Lundvall B.-A. and Borrás, S. (2005) *Science, Technology, and Innovation Policy, The Oxford Handbook of Innovation*, Oxford University Press

Lundvall, B-Å and Johnson B. (1994) The learning economy, *Journal of industry studies,* 1(2), 23-42

Lundvall, B.-Å. and Tomlinson, M. (2002) "International benchmarking as a policy learning tool", In Maria Joao Rodriguez (ed.), *The new knowledge economy in Europe: A strategy for international competitiveness with social cohesion.* Cheltenham: Edward Elgar

Lundvall, B.-Å. and Tomlinson, M. (2000) "Learning b y comparing – reflections on the use and abuse of international benchmarking", In Sweeney, G. (ed.), *Innovation, Economic Progress and The Quality of Life*, pp. 120-136, Cheltenham: Edward Elgar

Maskell P. (1996) *Learning in the village economy of Denmark. The role of institutions and policy in sustaining competitiveness,* DRUID Working Paper Nr. 96-6

Rajsekhar, B, van Hulle, F, Jansen, J C (1999) "Indian wind energy programme: performance and future directions", *Energy policy*, 27, 669-678

Rosenberg, N (1982) *Inside the Black Box: Technology and Economics* (Cambridge University Press, Cambridge)

Sandelowski, M (1997) '"To be of use": enhancing the utility of qualitative research', *Nursing Outlook,* 45(3), 125-32

Sawin, J (2001) Doctoral dissertation, The Fletcher School of Law and diplomacy, Tufts University, Medford, MA

Singh, V (1998) "Wind Power Slows in India", *Clean Energy Finance*, 2(3)

Szabo, Vand Strang, V R (1997) 'Secondary analysis of qualitative data', *Advances in Nursing Science,* 20(2), 66-74

TERI Project Report No. 2000RT45, Survey of renewable energy in India, Tata Energy Research Institute, India

Thorne, S (1990) "Secondary Analysis in Qualitative Research: issues and implications", in Morse, J.M. (Ed.) *Critical Issues in Qualitative Research Methods. London: Sage.*

Windpower Monthly (WPM) 2004:25, "Denmark picked for global headquarters", *Windpower Monthly News Magazine A/S*, Denmark.

Wong S.F, (2005), "Obliging Institutions and Industry Evolution: A Comparative Study of the UK and German Wind Energy Industries", *Industry and Innovation*, Vol 12, No.1.

14

Wind Energy in India
The Future and the Challenge

A Lakshminarasimha and S Rath

The world urgently needs energy from non-fossil fuel sources and the best alternative has been wind energy which is totally environment friendly. Denmark is the first country to adopt wind energy as its prime source of energy. As a matter of fact, Denmark is the leading wind power nation in the world. Wind energy is green and clean, but it employs 20,000 people in the manufacturing of the equipment required. India too needs the help of wind energy and in an increasing quantum. The paper looks at the energy scene in India. It elaborates on the Indian wind energy industry and projects the future of wind energy in India. It also looks at the challenges to be faced when investment is planned in this source.

Introduction

Fossil fuels will not be available in abundance for ever. Predictions show that fossil fuels will start diminishing from 2025. Also, burning of fossil fuels is environmentally dangerous and is a threat to the health of the society.

Source: The Icfai Journal of Infrastructure, Vol. V, No. 4, 2007.

The major thrust to develop wind energy in India came in the early 1980s from the then Department of Non-conventional Energy Sources, now known as the Ministry of Non-conventional Energy Sources (MNES). MNES undertook an extensive study of the wind energy potential and installed a countrywide network of wind speed measurement stations. These have helped it immensely to assess the national wind potential and identify suitable areas for harnessing wind power for commercial use. The total potential for wind power in India is estimated at 45,000 MW. The total installed power capacity from all sources (thermal, hydro, nuclear, etc.) is 124,287 MW. Compared to this, the total estimated demand is around 2,30,000 MW. As such, a gap exists for utilization of all possible sources of energy.

Energy Scenario in India

India now ranks fifth in the world in total energy consumption and needs to accelerate the development of the sector to meet its future energy needs. The country has rich deposits in coal and is abundantly blessed with renewable energy sources such as solar, wind, hydro and bio-energy but has very small hydrocarbon reserves (0.4% of the world's reserve).

In the early stages of the wind power program, the Government of India established wind farms for the sole purpose of demonstration. These helped in creating an interest in this field. For the past two decades, government programs were responsible for the growth of wind power. Although there were many suppliers, power was supplied only to the National Grid.

Reform policies started by the government after 1992 helped the market expand in wind power development. After the initial phase of government-sponsored demonstration projects, the wind energy sector was liberalized and was opened up for private participation in 1992 supported by appropriate government policy incentives, fiscal incentives and other benefits. A host of incentives were offered to wind power plant entrepreneurs and willing companies, which included tax concessions (accelerated depreciation, tax holidays, excise and customs duty relief), soft loans and liberalized foreign investment procedures.

In the very first project commissioning, the most attractive fiscal incentive was 100% accelerated depreciation on wind power equipment. As a result, the Indian electricity sector changed the competitive advantage of wind power market. This generated a significant demand for wind power in the private sector. Banking and foreign exchange reforms aided this source of electricity as it seemed feasible for the companies to reap benefits. Technology transfer occurred in this sector, as many international wind turbine companies backed by the governments of their home countries helped in supplying technology, and some Indian companies also formed joint ventures with wind energy equipment suppliers. Due to the implementation of these technologies, the costs of wind power and manufacturing were significantly lowered. In this adoption of technology, the government acted as a catalyst. At present the country has already developed a strong indigenous manufacturing base for wind power equipments with almost 80% of the equipments being manufactured indigenously.

The growth rate in wind power capacity which was at its peak in 1995/96 decreased thereafter. After the surge in growth that pushed India's energy utilization to the world's third spot, the investment fell sharply from mid-1996 to the end of 1998. This was due to the lowering of tax-credit incentives/benefits due to the implementation of Minimum Alternate Tax (MAT), reduction of corporate income tax by the Union Government and non-allowance of third party sales of electricity in some states.

The Indian Wind Energy Industry thus Far

India is one of the few countries in the developing world which actually pioneered the development of renewable energy. From the 1970s onwards India recognized the importance of natural sources of energy after which the sector witnessed a slow but steady growth for the next three decades. After that, there have been many developments which can be summarized as follows:

- 1981 – Commission for Additional Sources of Energy (CASE) was set-up;
- 1982 – A global first Ministry of Energy i.e., Department of Non-Conventional Energy Sources was established;
- 1985 – Installation of the first grid-connected wind turbine;
- 1986 – Beginning of the demonstration programme by DNES;
- 1987 —Setting up of IREDA to finance wind power and renewables;

- 1992 – The department was converted into a full-fledged MNES;
- 1992 – Recognition of renewable technologies for power generation in 1992 by their inclusion in the Eighth Five Year Plan (1992-97);
- 1992 – Policy to encourage private sector investments in power sector;
  1993 – Guidelines for wind power tariff by MNES;
- 1998 – Establishment of Centre for Wind Energy Technology (CWET) for testing and certification of wind turbines, fiscal and promotional Incentives: (*http://mnes.nic.in/wp4.htm*);
- Concessional import duty on specified wind turbine parts;
- An accelerated depreciation of 80% in the first year;
- Sales tax, excise duty reliefs;
- Loans through Indian Renewable Energy Development Agency (IREDA);
- Income Tax holiday; and
- Nine states have introduced these policies.

MNES Guidelines on Policy

- Wheeling charges of 2% (the charges to be paid to a Central Electricity Authority when energy is transferred from one grid to another);
- Third party sales at remunerative prices;
- Buy back facility at minimum price of Rs.2.25 per KWh for the base year 1994-95; and
- An annual escalation of 5% in tariff.

Future of Wind Power in India

India's achievements in wind power are definitely laudable if one considers the fact that it is the private sector that has been responsible for the growth. The wind power industry can get the boost only when suitably aided by incentives to make the projects operational in the Indian context. Wind power is generally subsidy-driven and will continue to get special attention including financial compensation due to the environmental benefits provided by the industry. The wind farm installations cannot be compared to the conventional power, as the plant load factor of a wind farm averages from 20%-30%.

Today, we are in a period that is similar to 1994 when this sector saw many new entrants (Figure 1). Recently the Reliance Anil Dhirubhai Ambani Group took over NEPC. Even many public sector companies such as Oil and Natural Gas Corporation (ONGC), Gujarat State Financial Corporation (GSFC), Minerals and Metals Trading Corporation (MMTC) and Hindustan Petroleum Corporation Limited (HPCL) are vying to get into this industry. All this has been possible because of the newly established regulatory environment. It has acted as a catalyst for the growth of the industry as experienced so far.

Figure 1: Wind Power Development in India

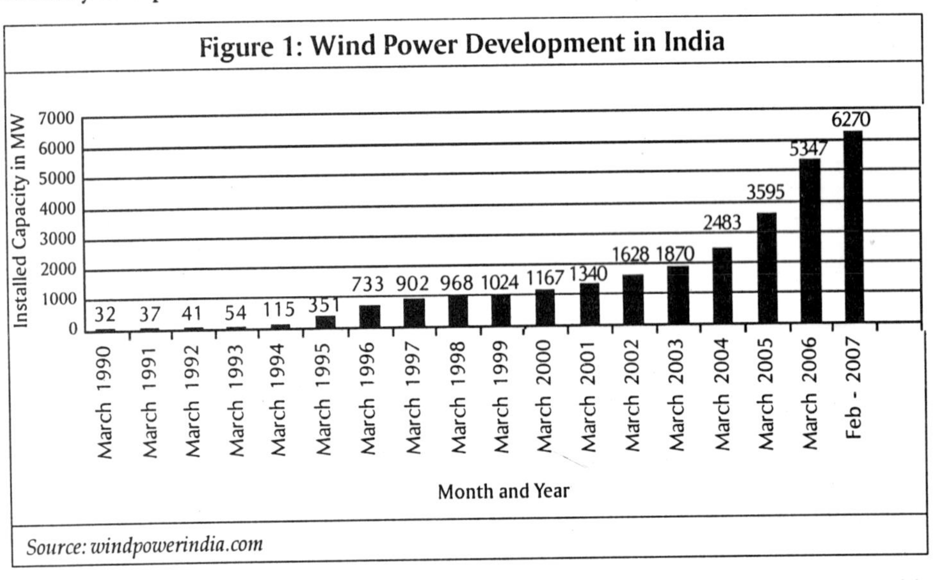

Source: windpowerindia.com

Wind Energy has now established itself as a source of electricity all over the world. While wind turbines in the range of 1MW-4MW are being set-up globally, in India all machines are less than 2MW capacity. When 55KW wind turbines were set up in India, it was 200kW-600kW capacity wind turbines in Europe; and when it was 500kW wind turbines in India, it was 1MW-class wind turbines in Europe.

Offshore projects which are being set up on a large scale in different parts of the world will also be starting in India in near future. With offshore projects potential in the multi-megawatt class to be assessed, there would be an upward revision of the 45,000MW wind energy potential in India.

The intermittent nature of the wind power has been an issue for years. However, keeping in view the Research and Development (R&D) that is happening all over the world, this problem could be overcome by implementation of new techniques. Better forecasting techniques are being evolved to predict the output from a wind farm over the next 12 hours. These are self-learning computer models that improve with use. This will enable the grid operator to plan for any shortfall or increase in generation. R&D on large storage devices may ultimately enable a wind turbine to store energy for half an hour or so, which would help the wind turbine give a steady output.

Wind power investors would basically be interested in supplies from reputable manufacturers who can offer consistently good support services in respect of operations and maintenance even after the warranty period of the equipment is over. The cost of wind power projects being predominantly decided by turbine cost, the lower the cost of the turbine, the lower the project cost which serves as an impetus to the investor.

Fiscal and Financial Incentives: The Current Scenario

To promote commercial development of wind power sector, the government has in place a package of fiscal incentives which include accelerated depreciation benefits, tax holidays, relief from customs and excise duty and liberalized foreign investment procedures. Accelerated depreciation to the tune of 80% on wind energy devices/ systems is allowed in the first year of the installation of the project. Income tax benefits under Section 80 I-A of the Income Tax Act etc. are also important to an investing company investing in a wind power project.

The Electricity Act, 2003 provides for the establishment of State Electricity Regulatory Commissions, which in turn, would mandate a minimum percentage of power to be purchased from renewable sources of energy by the power off-takers in the state.

Challenges Faced by the Indian Wind Energy Sector

The wind energy sector in India, despite its phenomenal growth, has seen many ups and downs and even today the growth is hampered by constraints. The growth and geographical spread of commercial wind power projects in the country can be directly related to the availability of conducive policies for setting up such projects and for policies pertaining to purchase of power generated. If wind farm locations are mapped in the country, it would be clearly seen that almost all installations are located in states that had announced long-term conducive policies relating to power tariffs, wheeling and banking charges (Figure 2).

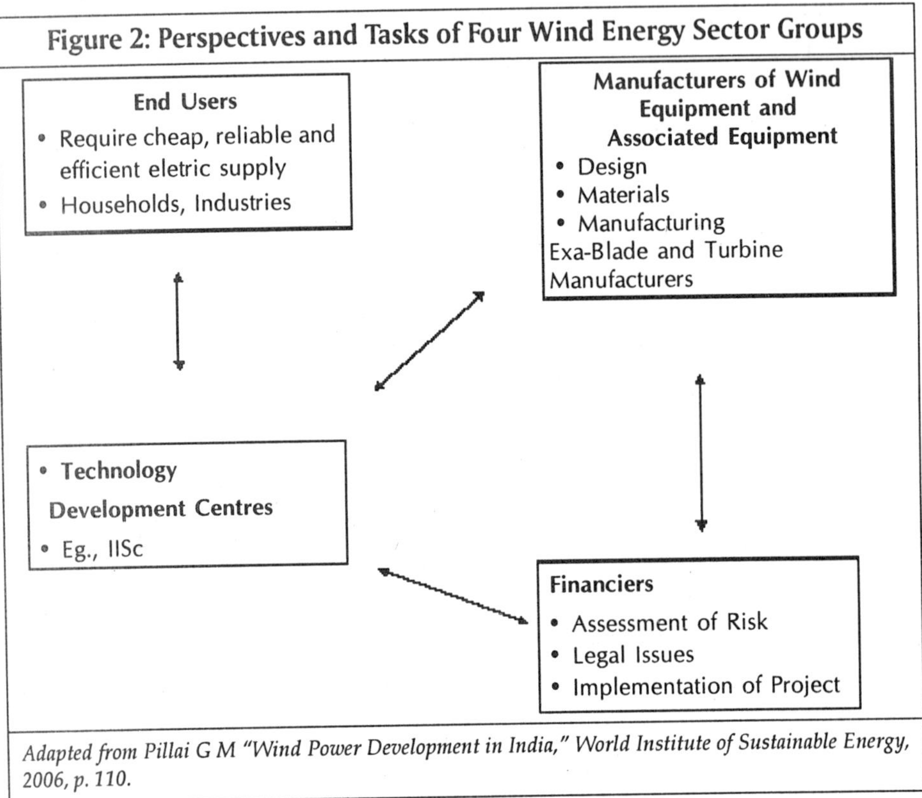

Figure 2: Perspectives and Tasks of Four Wind Energy Sector Groups

Adapted from Pillai G M "Wind Power Development in India," World Institute of Sustainable Energy, 2006, p. 110.

- Wrong location has been one of the greatest limitations to the development of wind energy projects. Most of the potential wind power sites are located in areas where infrastructural facilities are virtually non-existent. Suitable roads have to be built to transport the long blades and heavy equipment;
- Transmission lines and power evacuation facilities also have to be set up by wind farm developer and thus it places an undue financial burden on the project cost, which may lead to lower financial viability; and

- Electricity generation from wind is infirm and suffers from the intrinsic disadvantage that the wind electric generator produces power only when there is sufficient wind and not necessarily when there is demand for power.

Next Steps for India

- **Proper Tariff Structuring and Inflation Adjustment:** In most European countries the feed – in tariff legislations have been enacted to ensure renewable support price for renewable power;
- **Renewable Portfolio Standards:** Internationally, Renewable Portfolio Standards (RPS) are actually signified as a market driven mechanism that makes sure that a certain specified minimum percentage of energy in a system is generated from renewable energy resources and can be implemented as a policy in the state. Under Section 861(e) of the Electricity Act, 2003 at least nine states have initiated action to implement a quota system for purchase of renewable power by utilities;
- **Slashing Fossil Fuel Subsidies:** Rates of fossil fuels like diesel have been subsidized by the government. So, the government should take steps to slash those subsidies so that the wind power demand in the country is not affected. Transition to sustainable energy urgently requires a gradual and planned phasing out of these thousands of crores of rupees. They should instead be deployed for developing new energy technologies;
- **Innovative Financing:** The government should also look into the financing of the wind energy projects which should include new policies which actually would help the potential wind energy promoters. The investment in wind power in India was about Rs.8,500-Rs.9,000 crore, out of which the loan component was Rs.7,000 crore. This is bound to grow at a much faster rate in the coming years. These financing mechanisms should act as a catalyst for the growth through large-scale involvement of the commercial banking sector;
- **Public-Private Partnership in R&D:** The emerging trend these days are the public-private partnership where most of the wind turbine related research work is currently going on. The Ministry of Non-conventional Energy Sources may consider initiating a well-coordinated string of projects involving the government-funded premier technical institutions like the Indian Institute of

Technology (IIT), National Institute of Technology (NIT) and major engineering colleges;

- **Study the Offshore Potential:** Offshore wind has emerged as a high-growth area in Europe and the US where new technology development efforts have reaped in better prospects in this sector. India is yet to assess its offshore wind potential in its long coastline thousands of kilometers long. This sector will provide the much-needed market enlargement and will also attract major investments; and
- **Production Tax Credit:** At present, the wind power enjoys 80% accelerated depreciation on capital investment. A better proposition which could be followed on the global lines is to link income tax incentive to generation instead of capital costs.

Barriers to Accelerate the Growth of Wind Energy in India

- Inherent, intermittent nature of renewable energy sources leading to relatively lower capacity utilization factors;
- Relatively high capital costs compared to conventional power systems. A wind power plant of 1MW costs Rs.4.5-Rs.5 crore to setup, whereas a similar set-up would cost Rs.1-2 crore;
- Inadequacy of technology development research on wind energy;
- Relatively low understanding or confidence level of the investors, financial institutions, decision makers about the weaknesses, threats impeding wind energy projects and opportunities; and,
- Absence of a green power market or commercial orientation of policies such as preferential tariffs, necessitating consistent fiscal/financial concessions for making wind energy a commercially attractive proposition.

Improving Energy Efficiency

Indian industries do not give major importance to energy saving. The three main reasons for high energy consumption by Indian industry are:

- Most of the manufacturing units still depend on old machinery;
- The relatively high cost of capital as compared to European/US Standards; and
- Uncertainty about the long-term growth of the particular industrial sector.

According to a report, the World Bank indicated that the Indian industry can save 20%-30% of total energy consumption. In India this sector has serious problems such as shortage of supply, lack of investment for additional capacity and the problems with the coal based power plants due to their emissions which cause environmental problems, that's why there is a need for energy conservation and efficiency improvement in the Indian power sector. The high auxiliary consumption and transmission and distribution loss further worsens the problem.

According to several studies, it has been estimated that nearly 30,000 MW can be saved with the energy conservation programs. Studies and experience have indicated that most of India's megawatt potential can be captured at substantially lower costs compared to the cost of capacity additions, which currently stands at over US$1 million per MW. However, inspite of good returns and short payback periods for energy efficiency investments, most of India's end-use energy efficiency potential remains largely untapped.

Future Wind Potential in India (The Projections)

In this analysis the variables taken are with their importance or their influence to the growth of wind industry in India. The variables are as follows:

- **GDP:** The GDP (Gross Domestic Product) of the country is taken into account because this factor is mainly responsible for the overall growth of different sectors in the Indian Economy. The increase in the growth rate of the GDP will fuel the demand for the electricity from the wind sources.
- **CUF:** The CUF (Capacity Utilization Factor) is a ratio of the actual energy produced in a given period to the hypothetical maximum possible, i.e., that produced by a turbine running full time at its rated power. The CUF depends on several factors such as wind regime of the place, the quality and age of machines installed etc.

- **Tariff:** Calculation of the cost of electricity can be expressed as a ratio of the total annualized costs to the total electricity (KWh) generated. The tariff is also a major factor which contributes to the growth of the wind energy sector in India.

- **Project Cost:** Project cost is the cost incurred by the wind energy promoters in setting up of the wind farms. The costs include the cost of the turbines, blades, mast and the nacelle which are the major components of a wind mill.

- **Outlay:** The outlay is the government's total investments in a particular sector where it invests or promotes the industry. The more the amount of outlay, the more the amount of interest generated by the companies in setting up of the wind farms.

- **Size of the Rotor:** Size of the rotor is a technological variable which shows the impact of technology on the productivity of the wind energy installations. From units of 20-60 KW in the early 1980s, with rotor diameters of around 20 m, generators of single wind turbines have increased to 5,000 KW with rotor diameters of over 100 m. This is also a major factor for the overall increase in the production/installations (Figure 3).

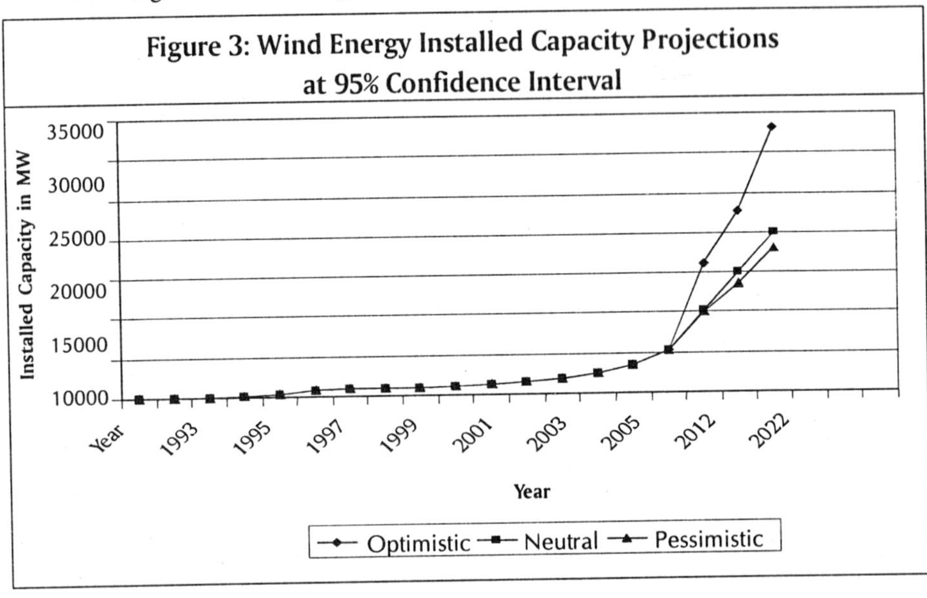

Figure 3: Wind Energy Installed Capacity Projections at 95% Confidence Interval

By considering these above mentioned factors in the analysis three different scenarios are prepared which are based on the macro-economic and internal factors that are mainly responsible for the growth of wind power in the years to come. The three scenarios are as follows:

- Optimistic
- Neutral
- Pessimistic

Optimistic: The most ambitious scenario follows a development path which would give maximum outcome/projections by considering the above factors in calculating the projections for the 11th, 12th and 13th Five Year Plans. The assumption here is that all the policy options work in favor of renewable energy, country's economic factors and the technology which actually affects the growth of wind energy sector in India (Table 1).

Table 1: Optimistic Scenario Projections

Year	GDP	Inflation	CUF	Tariff	Project Cost (in lakhs)	Size of Rotors	Outlay
2012	9.5	4	25	3.75	400	120	1700
2017	9.5	4	30	3.75	350	160	2200
2022	9.5	4	35	4.00	300	250	4000

Neutral: This is the most conservative scenario which is based on the current economic and technological trends that affect the growth of wind energy market in India (Table 2). The projections of the 12th and 13th Five Year Plan are calculated by increasing the projection of the 11th Five Year Plan by 100% in each plan.

Table 2: Neutral Scenario Projections

Year	GDP	Inflation	CUF	Tariff	Project Cost (in lakhs)	Size of Rotors	Outlay
2012	8.4	5.5	20	3.5	450	80	1350

Table 3: Pessimistic Scenario Projections							
Year	GDP	Inflation	CUF	Tariff	Project Cost (in lakhs)	Size of Rotors	Outlay
2012	6.0	6.5	20	3.50	450	80	1000
2017	7.0	5.5	22	3.75	400	100	2500
2022	7.5	5.0	25	3.75	400	150	2500

Pessimistic: In this scenario the projection is based on the least increase in the factors responsible for the growth of wind energy market in India (Table 3).

The Result

Table 4: Capacity Additions in the Wind Energy Sector – Five Year Plan-wise			
Plan	Optimistic	Neutral	Pessimistic
11th	10850	4865	4690
12th	17409	9730	8280
13th	27841	14590	12700

By using the multiple regression analysis in SPSS the following projections for the 11th, 12th and 13th Five Year Plans are arrived at Table 4. These are based on the factors shown above.

Conclusion

Energy demand and supply gap is widening day by day. If planned now and executed carefully, we can actually get more energy efficiency and new renewable technologies which will definitely take care of the future surge in demand.

Wind power is the fastest growing renewable energy in the world and to capitalize on this advantage all-out efforts should be made to promote and fully develop this energy on a large scale and to explore new avenues.

The projections provided would definitely help the upcoming wind energy promoters who are vying to get into this field. The challenges faced by the industry including location and transmission infrastructure costs, while taking into account the non-alignment between the demand need time and the electricity supply available time when it is based on wind energy, needs also be considered.

(A Lakshminarasimha is a Faculty Member at the Icfai Business School, Bangalore. He can be reached at lnarsimha@ibsindia.org

S Rath is a Student of PGDBA at the Icfai Business School, Bangalore. He can be reached at swagat03_oec@yahoo.com).

References

1. *http://euindiawind.net/pdf/eiwen_documents.pdf*
2. *http://google.com*
3. *http://mnes.nic.in/wp4.htm*
4. *http://nationalwind.org/publications/statepolicy/chapter4.pdf*
5. *http://planningcommission.nic.in/reports/genrep/rep_intengy.pdf*
6. *http://wikipedia.org*
7. *http://www.gwec.net/fileadmin/documents/Publications/GWEC_A4_0609_ English.pdf*
8. *http://www.mnes.nic.in*
9. *http://www.ucsusa.org/clean_energy/clean_energy_policies/production-tax-credit-for-renewable-energy.html*
10. *http://www.wind-watch.org/*
11. Pillai G M (2006), "Wind Power Development in India," World Institute of Sustainable Energy, p. 110.
12. *www.cwet.tn.nic.in*
13. *www.gwec.net*
14. *www.iea.org*
15. *www.iredaltd.com*
16. *www.windpower.com*

15

Research and Assessment of Wind Turbine's Noise in Vydmantai

Bronius Jaskelevičius and Natalija Užpelkienė

The decreasing reserve and growing costs of coal, oil and gas has persistently led us to use renewable energy resources. Using them, environmental harm, air pollution and greenhouse effect can significantly be decreased. According to the European Union Parliament's and Council Directive 2001/77/EC approved in July 27, 2001, the member countries (Lithuania has been a member country from May 1, 2004) must seek to use not less than 12 per cent of renewable energy sources in total energy balance and 22,1 per cent in the country's energy balance (including large hydroelectric power station). The numbers are significant. Thus, using wind energy can be one of the ways to fulfil the requirements. Lithuanian seaside, where the strongest winds are blowing, has been the most favourable place for the wind turbine's instaliation. However, usage and development of wind energy has been facing resistance from the society. One of the biggest arguments against the wind turbine is the noise which can cause inconveniences for the permanent residents

Source: www.vgtu.lt/english/editions/ Originally published: Journal of Environmental Engineering and Landscape Management, 2008, 16(2): 76-82. http://www.jeelm.vgtu.lt/upload/environ_zurn/2008_2_76-82-jeelm-2008-2-jask.pdf

and holidaymakers of the health-resort zone. Therefore, environmental impact of the wind turbine requires exhaustive investigations. The research was carried out in the vicinity of the only one presently active industrial wind turbine in Vydmantai (region of Klaip5da, Lithuania). During the investigation, it was established that the highest noise level was registered opposite the airscrew. The smallest noise level was measured in both sides from the tower in windscrew plane. It has been measured that noise level is noticeably higher in Vydmantai wind turbine's zone than in the background, which does exceed 250-300 m (5-6 windscrew diameters).

1. Introduction

Fast improvement of technologies, an increasing number of transport means, growth of people's living standard have been observed all over the world. Because of increasing electric energy consumption and intensive usage of natural resources environmental pollution is becoming much higher. If measures of pollution diminution and resources sustainability development are not taken into account, it is obvious, that the global catastrophe will become real. The results correction will require a few mankind generations.

Lately renewable energy resources are being used more widely than before. Wind energy is one of the most acceptable and significant natural resources. Although, with the increasing interest, distrust and causion are rising as well, caused by the wind turbine's environmental impact. These problems are noise, airscrew sparkling and vibration. These factors may have negative effect on people, pose a danger to flying birds, and harm local soil biocenoses. Besides ecological problem caused by the wind turbines, the landscape's view can be disrupted.

Various technical, energy and economic wind turbine's aspects are often analysed in scientific literature. However, the noise, which is the main society disagreement argument, has not been widely emphasised (Petrauskas 2001; Petrauskas, Adomavicius 2001). To fill this gap, wind turbine's noise in Vydmantai (region of Kretinga) was examined. The conclusions about wind turbine's noise levels were made after the measurements and generalisation of research results.

2. Possibilities of Wind Energy Usage in Lithuania

In recent years wind energy usage has become very intensive in many European countries. Germany takes the precedence of wind turbines (WT), where installed power at the end of 2002 reached more than 12000 MW, Spain took second place – 4 830 MW, Denmark reached 2 880 MW, etc. First place among the Baltic states belongs to Latvia where installed power has exceeded 22 MW during the last years. The stimulus to use wind energy in Latvia is a high purchasing price of the wind electric energy – 44 ct/kWh (Katinas 2003). For example, in Lithuania the price is 22 ct/kWh.

For centuries Lithuania has been a country of winds. Already in the 16th century wind energy was used for grain milling and boards cutting in the whole Lithuania and especially in the seaside region. Only during the Soviet period energy sector became centralised in Lithuania.

The main cost part (82 per cent) of electric energy is spent on energy transmission from the country's eastern power plants to consumers. Electric energy in 400 kilometres way undergoes a lot of losses. Bringing electricity manufacturing closer to the consumers, supply distance becomes shorter and the losses of electricity are smaller. Therefore, decentralising renewable energy sources will be very cost-effective. Nowadays, like in 16th century, nearly every small town can be a wind turbine. The costs of electricity could lessen twice (in recent prices it is 14 cents per kWh). For consumers electricity prices will not rise taking into account today's wind turbine energy price which is 22 ct per 1 kWh. Because of constant improvements in the wind turbine's construction and manufacturing, wind energy cost has been going down and has become competitive with the price of mined fuel (Markevicius and Katinas 2003). Until the year 2010 200 MW sum power park of wind turbines will be built at the seaside. It will be enough to satisfy Klaipėda's needs for electric energy.

Until the year 2020 country's need for natural gas can be decreased by 50 per cent only by using the wind energy. The remaining 50% will be acquired from the solar, hydrogen and other renewable energy sources. This is a broad ground for scientific investigations and innovation using European Union's financial support (Paulauskas 2006). In the research done, it was predicted that the highest electric energy increase would be reached by development of the wind energetics, for example, ~0 MW in the year 2002, 170 MW in 2010, 500 MW in 2020 (Jankauskas 2004). Because of mankind's uncontrolled activities global climate warming is observed. It is a serious warning, which must be taken into account. According to the Kyoto protocol for European countries (for Lithuania as well) the task is to decrease the emission of carbon dioxide (CO2) by 8 per cent in comparison with the level of the year 1990. In Lituanian National Energy Strategy it also is foreseen to use more natural gas and more pure fuel instead of mazute (Burneikis 2002).

Wind energy development has started only in recent years. One of the first works about wind energy possibilities in Lithuania was research done by Lithuanian Energy Institute's scientists (Katinas and Tumosa 1995). Nevertheless, a part of society and some of its representatives individually are against wind energy, treat it suspiciously or indifferently.

During the first wind turbine construction at the Lithuanian seaside society testing was carried out. The results showed that more than half a thousand inhabitants (73 per cent respondents) who responded to the question approved the development of wind energetics, positively and only 6.5 per cent showed their disapproval (*www.takas.lt*, 29-04-2003). Even though, the results of the questioning are favourable, wind turbine installation process is complicated.

Because of insufficient information about this modern energy producer for Lithuania, 20 per cent of the examinees confessed knowing nothing about wind energy, its advantages and the modern technologies used.

During wind power plant building in Vydmantai the local press wrote about the hazards of the wind turbine's noise, landscape pollution, suffering birds and animals, irritation to people caused by shadows twinkling and electromagnetic disturbance (Gabartas 2005).

Therefore, during the development of the wind energy sector and taking into account the society's resistance, it is necessary to explain widely economic and environmental advantages, develop a reliable legal base, carry out necessary and exhaustive scientific investigations connected with wind power plants rational spreading out. The research results and their analysis must be accessible to everyone. Public opinion is very important and must be taken into account.

3. Investigation Methodology

There are two potential noise sources in the wind power plant, i.e., turbine blades, flying (crossing) the air, and a cabin with a transmission box and a generator. Constructors have paid a special attention to the turbine blades noise which was minimised after their construction improvement and applying special materials. Standing near the turbine, rotating blade's whizzle as well as transmission box and generator's noise can usually be heard. Nevertheless, moving away (increasing distance) from the power plant this mechanical noise is decreasing (or dying away).

Wind power plant's raised noise distribution, according to its nature, is shown in Figure 1 (Shen, Sorensen 2001).

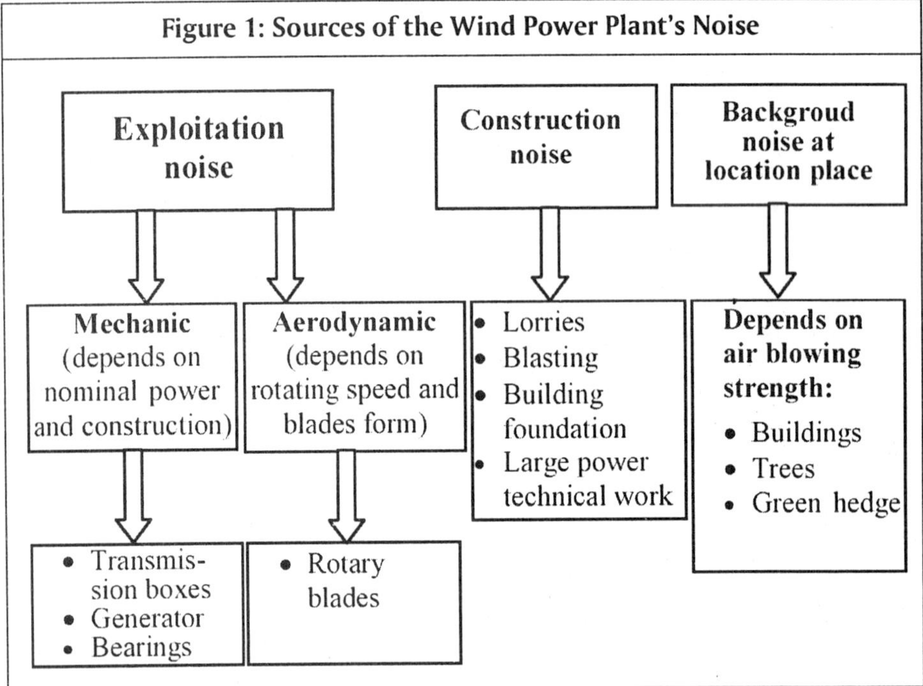

Figure 1: Sources of the Wind Power Plant's Noise

After the improvement of the wind power plant's construction and manufacturing technologies, level of noise of the turbine has been reduced and will further decrease. For example, constructors changed blade thickness (made them thinner) and projected (built) turbines "opposite to the wind" instead of "against the wind", i.e. wind blows first of all to the rotary blade, then to the tower. Because of this innovation, aerodynamic noise has significantly been decreased.

Today the main wind turbine's noise source is produced by turbine's mechanic components (rotating parts, rotary). Low frequency noise has decreased significantly when transmission box was extracted because generators with a rotary were directly connected to the airscrew.

In the near future new-type wind power plants will be built, where noise in 300 meters distance from the plant will not exceed library noise level.

More exhaustive wind power plant's noise formation mechanism is described below. According to its construction, noise of the wind turbines is divided into two groups (McKenzie 2000):

- mechanic;
- aerodynamic.

Mechanic noise: Mechanic noise appears because of different movements of the contact of surfaces and mechanic dynamic phenomenon in these surfaces. The main mechanic noise sources of wind turbines are:

- support bearing of the airscrew axle;
- transmission box;
- generator;
- cabin rotating mechanism;
- cooling ventilator;
- other equipment (oil pumps, cooling devices compressors, serving motors, wind speed gauge, etc.).

Aerodynamic noise. Aerodynamic noise is divided into three main groups:

- low frequency noise;
- turbulent vertical noise;
- aerodynamic surfaces (profiles of blades) noise.

Low frequency noise is formed when airscrew blade is crossing different air speed streams and the latter appears near the tower (in a vertical plane). Air stream in this plane is driven by the wind and after reverberation to the tower air stream moving speed changes.

Turbulent vertical noise: When a solid body structure (in our case the airscrew blades) faces air turbulation streams, wide band noise named turbulent noise occurs. The level of this noise, which is heard by people like a wheeze, and spectric composition depend on airscrew blades' construction. The noise caused by rotating blade's edge and in this way "cutting" airstreams is called *own noise created by aerodynamic surface* (Wei 2005). It is formed when chaotically moving airstream is mixing because of the moving blade. Parameters of this noise depend on airstream speed, blade's profile, attack angle and blade's surface properties (especially smoothness). Spectrum of aerodynamic surface own noise is wide banded.

Measuring Methodology: For measuring air turbine's noise any general noise making source measuring methodology can be used. So it can also be applied for measuring and investigating of any other noise making equipment sources. The methodology corresponds with the hygiene standard HN 33-1:2003 (HN 33-1:2003), of the Republic of Lithuania Metrology Law (Žin. 1996, No. 74-1768) and the European Union legal requirements.

Air turbine's noise has to be measured when the wind speed is not less than 6–8 m/s. During the measurement, microphone has to be covered by a special protecting screen and turned to the noise source, i.e., to the wind turbine. It is forbidden to carry out measurements during the rain, snow or fog. The measuring device microphone has to be at the height of 1,2–1,5 m from the ground level.

Besides, a frequency characteristic filter must be used for the measuring, because wind turbine's noise intensity is small (< 80 dB) and the filter sensitivity dependence is higher for small intensity noises than C characteristics filter.

Investigated Object: The only active industrial wind turbine in Lithuania, of 630 kW power, was chosen for the investigation. The vicinity is a small town Vydmantai. The plant is equipped with rather up-to-date turbine produced by a German company "Enercon GmbH", which makes a very popular wind power plant E-40. There are more than 3500 wind turbines of this type in the world.

Its short characteristic is presented in Table 1.

Table 1: Short Characteristics of the Investigated Wind Turbine (Nevardauskas 2005)

Tower Height	76 m	Cabin Mass	30 t
Windscrew axle equipment height	78 m	Minimum air speed	2,5–3,0 m/s
Airscrew diameter	43,3 m	Nominal power wind speed	11,5 m/s
Blade length	20,5 m	Maximum wind speed (electric resistance)	25 m/s
Maximum object height	99,5 m	Maximum wind speed as for the building (mechanical resistance)	52,5 m/s
Airscrew touched area	1472,5 m2	Single blade's weight	1,1 t.
Airscrew rotation speed	18–34,5 turns per min.	Linear speed of blades' end flight	41,0-78,6 m/s

A precision analyzer 2260, produced by a Danish company Bruel & Kjaer, has been used for noise measuring (Environmental Engineering 2005).

Wind turbine plant's noise measuring course. During the measurement of the wind power plant's noise, orientation of the airscrew was taken into account, mainly in the directions "opposite the wind", "before the wind", "in one plane with airscrew", and etc. It was done so, because the noise level depends on the observer's position with respect to the airscrew, which is always oriented to the wind.

Noise level was measured in selected distances – 25 m, 50 m, 160 m and 200 m from the tower. The directions were chosen every 45 degrees starting at the wind tower basis, i.e. opposite the wind (opposite the tower), before the wind (behind the turbine's power), from the left and from the right side with respect to the tower. Noise was measured in 36 points totally. 10 minute time was allotted for every point.

Measuring the environment and conditions: The wind turbine has been built not far from Palanga roundabout near a small city Vydmantai. There is a small forest nearby. This is why strong winds are always blowing in the territory. Vydmantai is rather far from a possible acoustic noise zone boundary.

It was sunny during the investigation and a medium strength wind was blowing. According to the data of Hydrometeorologic Service for the day of the examination, wind speed was 5,1 m/s (average) and 9 m/s (maximum) at the height of 10,5 m. At the height of the wind turbine's shaft (78 m) the wind blowing speed was 7-12 m/s.

4. Results and Analysis of Investigation

During the investigation of noise, produced by the wind power plant in Vydmantai, such parameters were measured:

1. Average equivalent noise level LAeq which is usually expressed in decibels (dB).
2. Minimum and maximum noise levels – LASmin and LASmaks and their difference PL.

Moreover, a special programmining equipment 2260 inside the noise analyser allowed to measure such parameters:

- average equivalent noise level distribution;
- spectric noise composition which is calculated by rapid Furye transformation.

Measuring results in the main directions are presented in Table 2.

Foreign scientists investigation conclusions (Figure 2) have been confirmed by the investigation results in Vydmantai (Figure 3).

It has been established that noise pressure level differs when observer's position is changing – when noise wave spreading direction is changed (Figure 3). The receiver around the wind power plant, the center of which was a wind turbine, was moving in circles, and the circle rays changes were from 60 to 200 m. Wind blowing direction was from 180 to 0 degrees. The examination results showed that noise spreading depends on wind turbine's rotary position in respect with the receiver and purposefulness diagram resembles bipolar noise source purposefulness diagram. Taking this into account it becomes clear, that the nearest dwelling-houses location in respect to the wind turbine is the best when we draw in noise source purposefulness diagram tangentic lines (tangents). Therefore, it is necessary to evaluate exact direction of the wind and its impact on the noise purposefulness diagram.

Table 2: Wind Power Plant's Average Equivalent Noise Level L_{Aeq} Measuring Results

Measuring Points	Wind Turbine is Average Equivalent Noise Level L_{Aeq}, dB				
Distance from the wind turbine, m	0	25	50	100	200
N (opposite the turbine's tower)	54,9	52,4	54,1	51,8	48,2
NE	--	57,5	59,4	53,3	55,8
E (from the right side)	43,2	46,6	47,4	46,2	40,4
SE	--	51,1	52,5	50,5	47,2
S (before the turbine's tower)	49,8	52,8	53,7	52,3	48,3
SW	--	51,5	52,3	50,1	48,5
W (from the left side)	42,7	49,5	47,6	46,4	43,3
NW	--	53,9	54,3	52,2	47,9

Figure 2: Noise Pressure Level in Different Points in Respect to the Observer

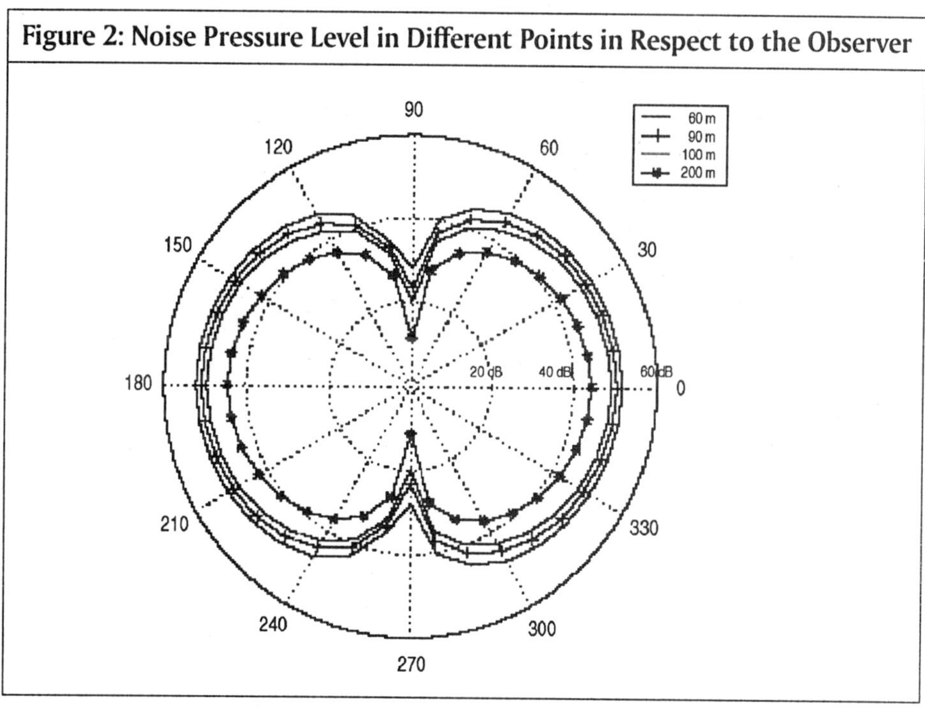

According to the wind turbine noise examining preliminary recommendations of the United Kingdom Economy and Industry Department, wind turbine's noise level has to maintain on average 5 decibels (A) difference of level between evening and

Figure 3: Noise Level Dependence on Measuring Point's Location in Respect to the Wind Power Plant's Orientation

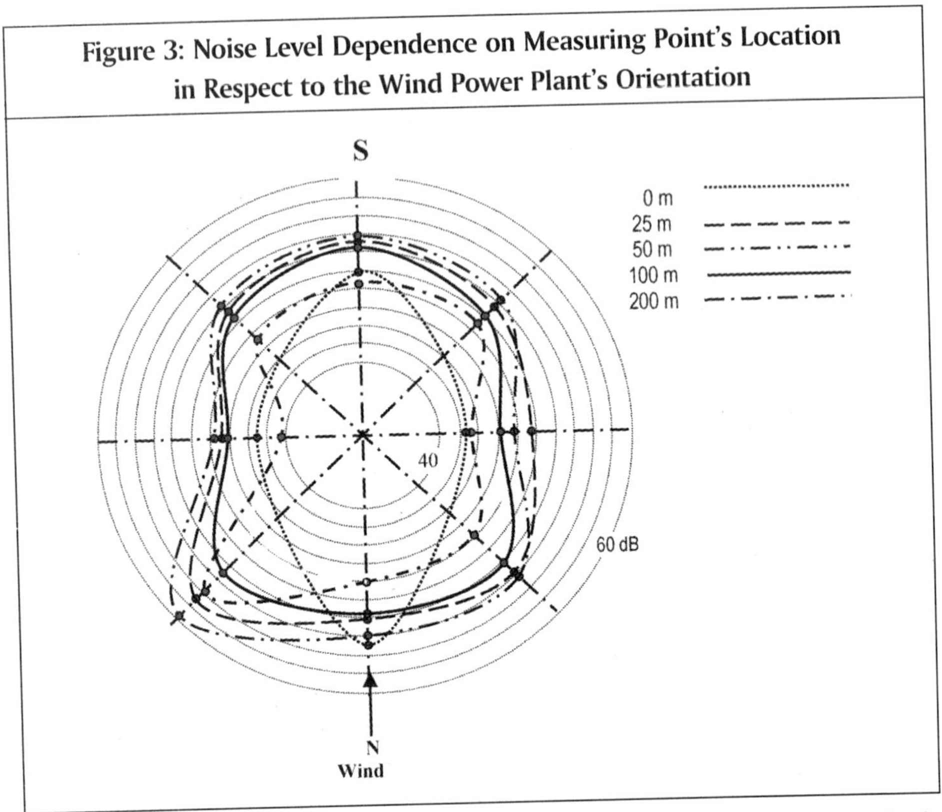

night time noise levels. Practically a lot of noise sources comply with these standards, except for the noise caused by transportation and building of the wind turbine, but these bigger noises usually are allowed, furthermore they are temporary. The lower noise level limit can be fixed between 35–40 dB (A).

Carrying out the measurements maximum wind turbines caused noise level was fixed L_{max} = 70,5 dB, on the other hand minimum fixed noise level during measuring time was equal to L_{min} = 38,1 dB.

During the analysis of the measured results it was noticed, that noise level caused by the wind power plant depends on the distance from noise sources. Maximum noise level was registered when a measuring device was 25-50 m away from the wind turbine basis (Figure 4 a, b, c, d).

Figure 4: Wind Turbine's Caused Noise Level Dependance on the Distance from the Noise Source (Allowed Noise Level 45 dB)

a – Opposite the wind power plant's tower

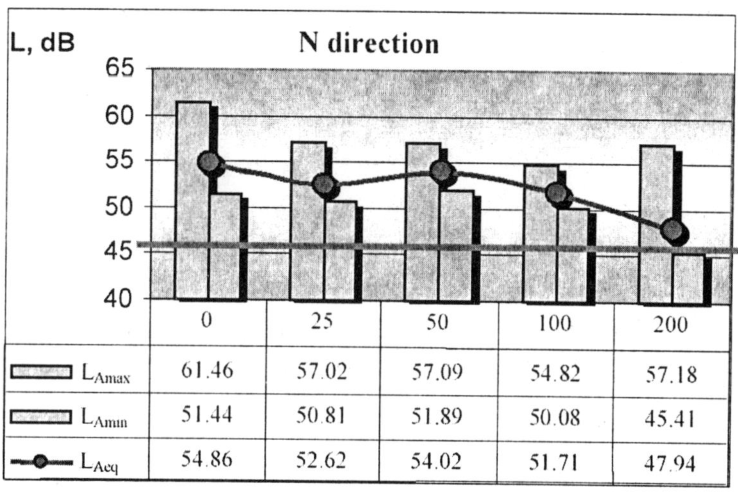

	0	25	50	100	200
L_{Amax}	61.46	57.02	57.09	54.82	57.18
L_{Amin}	51.44	50.81	51.89	50.08	45.41
L_{Aeq}	54.86	52.62	54.02	51.71	47.94

b – From the left side

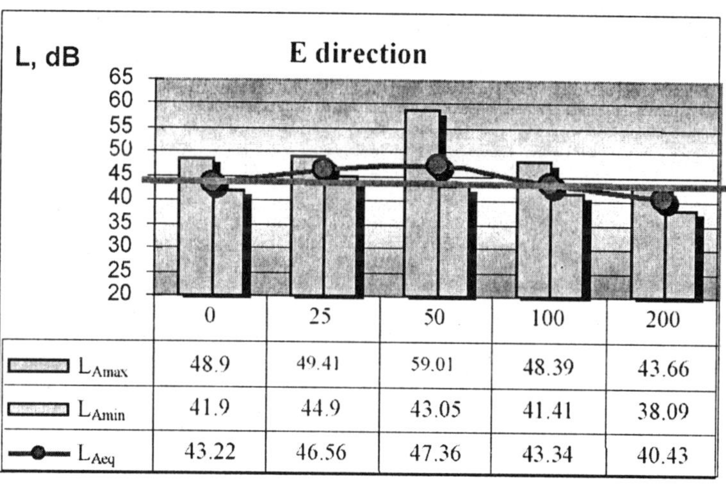

	0	25	50	100	200
L_{Amax}	48.9	49.41	59.01	48.39	43.66
L_{Amin}	41.9	44.9	43.05	41.41	38.09
L_{Aeq}	43.22	46.56	47.36	43.34	40.43

Contd...

Contd...

c – Before the wind power plant tower

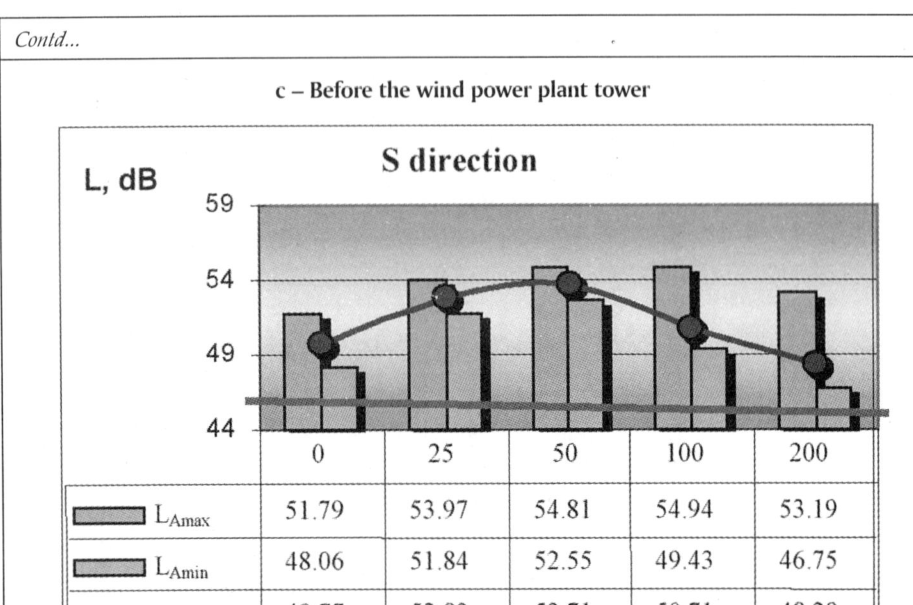

	0	25	50	100	200
L_{Amax}	51.79	53.97	54.81	54.94	53.19
L_{Amin}	48.06	51.84	52.55	49.43	46.75
L_{Aeq}	49.77	52.83	53.71	50.71	48.28

d – From the right side

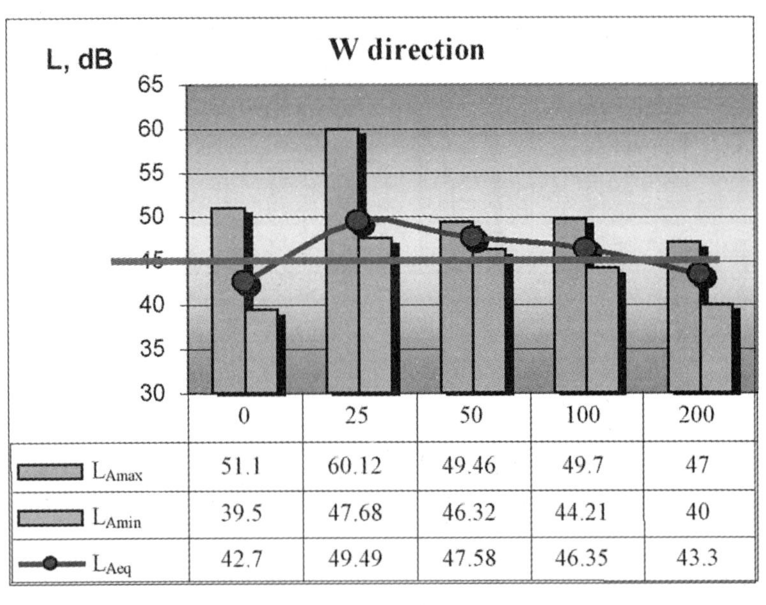

	0	25	50	100	200
L_{Amax}	51.1	60.12	49.46	49.7	47
L_{Amin}	39.5	47.68	46.32	44.21	40
L_{Aeq}	42.7	49.49	47.58	46.35	43.3

Analysing the wind turbine's noise spectric composition in 8 directions it was established, that two spectric components come to light near the wind turbine – low frequency (80-250 Hz) and high frequency (500-2000 Hz). Further from the wind turbine (1 > h) only high frequency noises ("wind wheese") remain. Attention must be paid to the spectric band of 500–2000 Hz which is clearly audible by people. However, no obvious dependence on any direction was observed. Therefore, an immediate conclusion can be made – that measuring noise spectric composition in any direction from the wind turbine, the results will be analogous.

Analysing the inconsistency of results some tendencies were observed. The biggest inconsistencies were noticed measuring the noise opposite the wind turbine's tower. The scattering was caused by the wind speed changes (gusty wind). The direction, which is opposite to screw orientation to the wind, distinguishes absolutely smallest discrepancy of the measured results. All in all two major causes were brought to light by the investigation:

1. The tower has influence on air stream movement, wind speed beyond the tower and air stream turbulent activity.
2. The tower is acting as a screen and decreases (obstructs) the noise formed before the tower. Therefore, the noise spreading is influenced by the air stream cutting with the wind screw. The noise is dispersed and only partially heard.

The noise spreading and the results are also influenced by other factors. Popular literature sources show that the terrain has a great impact on the noise level (Oškinis *et al.*, 2004).

It is necessary to notice that the increased acoustic level is one of the most fundamental wind power plant factors having an environmental impact and raising concerns of the permanent residents of Vydmantai. According to the valid hygiene standard HN 33-1:2003 Acoustic noise. Allowed levels in dwelling and industrial environment. General requirements of measuring methodology, the highest wind power plant average equivalent noise magnitude is 45 dB in a residential area, and in recreative zones the highest level must not exceed 40 dB. As one can see from the received results (Table 2 and Figures 4 a, b, c, d) in most cases this figure has been exceeded. However, this over fulfilment is best of all observed in radius of 25–50 metres around the wind turbine. Moving away from the wind turbine, the noise level noticeably decreases, however the standard (40–45 dB) is only reached in 200 metres from the turbine (Figure 4). These figures cannot be considered as final, because every turbine's acoustic indicators are summed up but another rules are valid for addition.

It has been proved that Lithuanian seaside is the most suitable place to develop wind energy. The fact, that it is a health resort, demands exceptional attention solving wind energy suitability for Lithuania.

To sum it all up it must be said, that science and advanced technologies, used in this area, can significantly decrease the impact of noise on the environment. A new wind power plant with vertical axle windscrew has already been made. Because of the up-to-date mechanisms, it does not cause sound noise (noise does not exceed background noise level) (Adomavicius, Balciunas 2003).

5. Conclusions

1. The main inconsistency of the measurement results was caused by the changing wind speed which during the measuring changed from 6 to 9 m/s.
2. The measured results scattering is proportional to the wind speed changing magnitude, i.e., the higher wind speed changing amplitude is, the higher measuring meaning's scattering is.
3. The wind power plant can be kept as a purposeful noise source which purposefulness coefficient D is 9,1 dB. The biggest noise level was observed standing opposite the airscrew. The lowest noise level was observed in both directions from the tower in the plane of the windscrew.
4. In the zone of Vydmantai wind power plant, which power is 630 kW and height is 76 meters, noise level is noticeably higher compared to the background noise. However, noise level in this zone does not exceed 250-300m, i.e. about 5-6 windscrew diameters.
5. Wind speed in the direction behind the wind turbine's tower is minimal because the tower obstructs the airstream movement. Therefore, turbulence is the lowest here.
6. The tower acting as a screen (obstructor) decreases formation of the noise which is still formed in front of the tower when the airscrew is "cutting" air stream flow.

(Bronius Jaskelevičius is Associate Professor, Department of Environmental Protection, Vilnius Gediminas technical University. The author can be reached at dtsas@ktl.mii.lt or aak@ap.vgtu.lt

Natalija Užpelkienė is Master student (environmental engineering), Department of Environmental Protection, Vilnius Gediminas Technical University (VGTU). The author can be reached at petras@erdves.lt).

References

Adomavicius, V.; Balciunas, P. 2003. Some possibilities of a small wind power plant efficiency increase, *Energetics* 1:38–45.

Burneikis, J. 2002. Energy and Environment. Available from Internet: *<http://ausis.gf.vu.lt/mg/nr/2002/03/03ener.html>*.

Environmental Noise. Brüel & Kjær. 2001. Brüel & Kjær, Sound & Vibration Measurement A/S. May 3, 2005. 53 p.

Gabartas, R. 2005. Alternative energy at a walk, Kauno diena 196 (17657), August 25. Available from Internet: *<http://www.kaunodiena.lt/lt/?id=6&aid=30703>*.

Hayes McKenzie. 2000. Noise from Wind Turbines. Available from Internet: *<http://www.britishwindenergy.co.uk/ref/noise.html>*.

Jankauskas, V. 2004. Electric energy production using renewable energetic sources support method, Energy 4: 1–11.

Katinas, V. 2003. Wind energy usage possibilities in Lithuania. Available from Internet: *<http://gstudija.tinklapis.lt/ltma/lindex3a.htm>*.

Katinas, V.; Tumosa, A. 1995. Wind energy usage possibilities in Lithuania. Vilnius: State enterprise "Energy Agency" Energy saving program direction. 37 p. Lithuanian hygiene standard HN 33-1:2003 Acoustic noise.

Allowed levels in dwelling and industrial environment. Measuring methodology and general requirements.

Markevicius, A.; Katinas, V. 2003. Wind energy development trends, *Energetics* 1: 22–27.

Nevardauskas, E. V. 2005. Demonstrative industrial wind power plant. Electric spaces. 56 p.

Oškinis, V.; Kinduryt5, R.; Butkus, D. 2004. The results of motor transport noise investigations on Vilnius–Kaunas–Klaip5da highrway, *Journal of Environmental Engineering and Landscape Management* 12(Suppl 1): 10–18.

Paulauskas, S. 2006. Sustainable energy. Lithuanian Wind Energy Association's view on country's energetic future. Available from Internet: *<http://www.eksponente.lt/darnioji_energetika.htm>*.

Petrauskas, G. 2001. Modern wind power plant properties and development perspectives, *Energetics* 1: 74–78.

Petrauskas, G.; Adomavicius, V. 2001. Wind energy resources and power plants technical economical indexes evaluation in designing stage, *Energetics* 2: 50–55.

Shen, W. Z.; Sorensen, J. N. 2001. Aerocoustic modeling of turbulent airfoil flows, *AIAA Journal* 39(6).

Wei Jun Zhu. 2005. Modeling of noise from wind turbines. Mechanical department, DTU. 105 p.

16

Wind Resources, Inc.

Rocky Higgins and Robert Keeley

This is a case study about the proposition to build and operate a 30 megawatt wind energy facility on the site of a company's easement to develop the Canyon Wind site located in the San Gorgonio Pass near Palm Springs, California. There emerged a difficulty as the private investors had encountered trouble in servicing the debt originally incurred to purchase the wind turbines. A continued use and upgradation of the existing wind energy equipment to much larger turbines was shown in market wisdom. The original agreement was about to expire and all the shareholders were anxious to cash out their investments as soon as possible. The CEO of the company was in a dilemma how best to capitalize on or develop this easement.

In July 2005, Mr. Charles Bittner, chief executive officer of Wind Resources, Inc. (WRI), needed to decide how best to capitalize on the company's easement to develop the Canyon Wind site located in the San Gorgonio Pass near Palm Springs, California. In 1985, WRI had secured the easement from the property's owner, the Bureau of Land Management (BLM), and entered into a complex 20-year agreement

Source: http://papers.ssrn.com © Robert C Higgins, Foster School of Business University of Washington, and Robert H Keeley, University of Colorado at Colorado Springs. Reprinted with permission.

with private investors, creditors, and Southern California Edison Company to build and operate a 30-megawatt wind energy facility on the site.

With the original agreement about to expire, Mr. Bittner needed to decide what to do with the easement. The site was reputed to be "one of the premier wind resources in North America" and conventional energy prices were rising sharply, so continued use of the site as a wind farm seemed an obvious choice. One option was to continue operating the site's existing but aging turbines. Ownership of the turbines and related equipment had recently reverted to WRI when private investors had encountered difficulty servicing the debt originally incurred to purchase them. A second option was to redevelop the site much as WRI had done in 1985 but this time using new, much larger turbines.

Recent increases in natural gas prices, the chief fuel used by electric utilities in the US, had stimulated considerable interest in alternative energy investments at noticeably higher prices, and Mr. Bittner sensed that WRI's principal shareholders were anxious to cash out their investment as soon as possible. But before he could decide what a reasonable selling price might be, and indeed before he could set a minimum price for any sale or auction, he needed to estimate the value of the project to new owners. (Exhibit 1 shows the Canyon Wind site and existing turbines. Exhibit 2 is a graph of natural gas prices over the past two decades, and Exhibit 3 records the volatility of gas prices over different time periods.)

The Industry

Today's wind energy business is a child of OPEC and Western governments. Concerned about American dependence on foreign oil and the environmental damage caused by use of fossil fuels, the US Congress passed the Public Utilities Regulatory Policies Act (PURPA) in 1978 as part of the National Energy Act. The legislation encouraged creation of energy from renewable sources, including wind power, and given certain conditions, required utilities to buy the energy at the utilities' highest "avoided cost." Avoided cost is the cost of the energy replaced by the renewable source.

Although PURPA is Federal Law, Congress delegated implementation to the states, resulting in a variety of regulatory schemes across states and the absence of any activity at all in others.

In 2004, California took the lead in PURPA enforcement when it required all major investment-grade utilities in the state to acquire one percent more of their power from renewable sources each year, so that by 2017 at least 20 percent of total electric supply is made up of renewable generation. It also mandated a bidding process requiring fixed-price 10 to 20 year contracts known as Power Purchase Agreements (PPAs) at prices based on the cost of new conventional generating sources. California's actions were largely in response to the devastating energy crisis it suffered in 2001 and a consequent desire to diversify supply, increase in-state production, and reduce reliance on natural gas-fired power.[1]

As further stimulus to alternative energy development, wind energy investors benefit from two lucrative tax breaks. Federal law allows owners to depreciate qualifying wind energy assets at an accelerated rate over a five-year period, even though the economic life of wind turbines and towers is closer to 20 years. In 1992 Congress created an annual Production Tax Credit (PTC) for wind and other renewable energy technologies. The credit is proportional to the energy produced and extends over the first 10 years of project life. The current PTC is 1.9 cents per kilowatt-hour with a cost of living adjustment of 2.5 percent a year. The original legislation was for only three years but has been renewed in fits and starts since. Current legislation extends the PTC through at least 2008.[2]

Wind power economics has improved dramatically over the past two decades, due primarily to the use of ever-larger turbines. The energy produced by a turbine is proportional to the cube of the wind speed and the square of the turbine's blade length. The gradual migration from turbines with blade diameters of 10 meters in the 1980s to diameters of 50 meters common in 2000 produced a 55-fold increase in power output, partly because the area swept by the blade is 25 times larger and partly because wind speed increases with blade altitude. Reflecting additional benefits of better turbine design, location, and computerized controls, the cost of wind-generated power has fallen some 90 percent in the past 20 years.[3]

---

1 "Overview of the California Model for Encouraging Renewable Energy Development," Thelen Reid Brown Raysman & Steiner LLP, Oil, *Gas and Energy Law Journal*, July 26, 2004. *www.constructionweblinks.com/resources/industry_reports_ne.*

2 "Congress Extends Wind Energy Production Tax Credit for an Additional Year," *American Wind Energy Association*, December 11, 2006.

3 "The Economics of Wind Energy," *American Wind Energy Association*, February 2005. *www.awea.org.*

Despite these improvements most wind power sources are still not competitive on price with fossil fuel power and may not be for years. According to data from the International Energy Agency (IEA) in Paris, the cost of electricity from coal-fired plants is 2.5-4 cents per kilowatt-hour, while the cost from natural-gas-fired plants is 4 to 6 cents. In contrast, energy costs from wind power range from 4 to 14 cents per kilowatt-hour, depending on size and location.[4]

Wind power accounted for only 0.5 percent of global and US domestic electricity production in 2004 according to the IEA. By 2030 the agency expects this figure to rise almost 7-fold to 3.4 percent. In the US capital spending on new wind projects in 2005 was on track to exceed $3 billion up from just $420 million in 2004. This would make wind power the second largest source of new electrical power for the year behind natural gas-fired plants.

Keys to a successful wind farm investment are a great site, an attractive long-term PPA, and continued government support of renewable energy resources. Wind farm investments require large initial capital outlays, followed by relatively stable long-term revenue streams. Because predicting wind velocity is much easier than predicting where new oil or gas reserves will be found, wind investments are considered safer technologically than conventional energy investments. The chief cost of a wind farm investment is the initial capital outlay, while the chief risks involve securing a favorable PPA and meeting a myriad of regulatory and permitting requirements, often including the placating of restive neighbors.

Wind Resources, Inc.

An experienced alternative energy entrepreneur founded WRI in 1985 to develop and market the Canyon Wind site located on BLM land. He designed the project to take full advantage of the liberal tax provisions available to qualifying renewable energy investments. As sponsor, WRI identified the site, negotiated a long-term, renewable development easement with the BLM, designed the wind farm, guided the project through a complex permitting process, secured a 20 year, fixed-price PPA with Southern California Edison, negotiated project financing, and identified a group of potential equity investors. With all the pieces in place, WRI then commissioned construction of the wind farm and sold the capital equipment and equity cash flow

4 "Renewable Power May Yet Yield Windfall," Keith Johnson, *Wall Street Journal*, p. A8, March 22, 2007.

rights for a period of twenty years to investors. (At the time target investors were wealthy individuals interested in available tax credits and shields. Tax laws changes in 1986 prohibited individuals from using tax shields generated on one activity to reduce tax obligations generated on another, so today's wind farm investors tend to be profitable corporations, such as General Electric, anxious to reduce taxes.)

WRI structured the equity transaction as an installment sale on the expectation that projected cash flows would service the installment debt. This feature was quite attractive to equity investors because it eliminated any initial investment on their part, guaranteeing there would be cash flow positive from day one. At worst, equity investors might default on the debt and have to walk away without the anticipated tax shields and profits, while on the upside, they would capture the anticipated tax benefits and any residual profits without any cash outlay.

WRI's profits would come from a sizeable development fee incorporated in the project's selling price, interest on the installment debt, a share of profits above a specified level, and annual fees for managing the facility. WRI also retained the right to repurchase the turbines at fair-market value at the end of the project's life in 2005 and to dispose of the site as they chose. Should they choose not to continue operating the property as a wind farm, WRI would incur a site restoration charge of as much as $2.5 million imposed by the BLM.

The 1985 Canyon Wind development did not live up to initial expectations chiefly because it never delivered more than 75 percent of targeted capacity. Inaccurate projections of wind velocity and persistence, combined with various unanticipated operating problems, were the chief contributors to the shortfall. Mr. Bittner was inclined to attribute these problems to industry growing pains that would not be repeated in any future redevelopment of the site.

In the end, equity investors received most of the anticipated tax shields but little in the way of profits. The situation was touch and go for a period when equity investors fell behind on installment payments to WRI, but they managed to recoup by the end of the period, in part by ceding ownership of the turbines and towers to WRI at the end of their contract. WRI's owners did better, receiving anticipated fees and interest other than shared profits, while retaining redevelopment rights.

The Alternatives

As the initial 20-year development contract approached maturity, Charles Bittner needed to recommend a course of action to his board. Growing dissention among WRI owners and financial problems at one inclined Mr. Bittner to rule out redevelopment by WRI. The possible imposition of a $2.5 million site restoration fee made abandoning the easement appear unattractive as well. Although other strategies were possible, Mr. Bittner decided to consider two in detail: continue to operate the existing turbines, or sell the BLM easement to another developer at auction.

Continue to Operate Existing Turbines

Exhibit 4 presents Mr. Bittner's analysis of the first option. Assuming WRI could keep the existing turbines operational for another 10 years by spending an additional $200,000 a year in current dollars on major maintenance, Mr. Bittner estimated the annual free cash flow from continued operation would be about $800,000 a year, for a present value of just over $4.7 million when discounted at ten percent. Ten percent reflected Mr. Bittner's understanding of industry practice when valuing unlevered wind energy cash flows.

Sell Easement to another Developer

Mr. Bittner reasoned that the highest price a wind farm developer would pay for the Canyon Wind site should equal the profit he could earn by redeveloping the site much as WRI had done in 1985. To help estimate the value of the site to a new developer, Mr. Bittner turned to MDS Energy Consulting, Inc., an experienced alternative energy consultant WRI had used in the past.

In their report, MDS identified seven milestones any redevelopment must achieve and briefly discussed the challenges to be addressed in meeting each.

1. **Site Control**. The current BLM easement expires in 2015 and needs to be extended before development can commence. MDS noted that obtaining an extension was likely but noted that the time, effort and expense involved could be "significant."
2. **Wind Resource Documentation**. WRI has twenty years of data on the strength and persistence of winds at the site. But use of much larger and taller turbines

means that additional data will need to be documented and confirmed as part of the redevelopment process. Efficiency is measured by a site's Net Capacity Factor (NCF), the ratio of the energy produced per year at a site divided by the theoretical maximum possible production.

3. **Regulatory and Permitting Approval.** Relevant county permitting requirements are some of the most highly developed and specific in the industry, which makes the permitting process time consuming, even if third parties do not oppose the project. Local residents immediately adjacent to the property had been quite vocal and effective in limiting efforts of other projects to develop nearby sites with newer and taller turbines. Moreover, because the site is on Federal property, significant environmental review might be required, including a new or updated Environmental Impact Statement. In MDS's view redevelopment permits could likely be secured but the outcome was not a foregone conclusion.

4. **Interconnection and Transmission Access.** The site has a working interconnection with the Southern California Edison grid, and it is likely this interconnection can be maintained and enhanced as necessary. The Federal Energy Regulatory Commission (FREC) must now approve applications for interconnection rights, and while approval appears assured, costs of enhancing the interconnection could exceed projections.

5. **A Long-term Power Purchase Agreement.** This is the lynch pin of any redevelopment. In order to secure necessary project financing, a long-term power purchase agreement with a credit-worthy investment grade (BBB- or better) buyer is necessary. MDS noted that the California Public Utilities Commission (CPUC) has recently determined that an appropriate price for renewable energy purchase under a 15-year PPA starting in 2005 should be $0.0588 per kWh. And while the CPUC's determination does not guaranty this price, it does provide a good indication of the potential market.

6. **Project Financing.** Once the redevelopment project has sufficiently documented its wind resource and secured site control, permits, an interconnection and a viable PPA, it must be financed. Under the current wind industry paradigm, the wind project owner must have a substantial appetite for tax credits. Leveraged after-tax internal rates of return (IRRs) in

the current market were typically in the mid-teens, while unleveraged IRRs were in the range of 10 percent. Interest costs on debt financing were 1.5% to 2.5% over 3-month LIBOR, and the first-year interest-coverage ratio had to equal at least 1.7 times.

7. **Project Construction.** The Canyon site has several characteristics that make it a challenging site for a modern wind energy development, including difficult terrain and access to the site. Hauling new, large turbines up and down the winding access roads may present a challenge. Ironically, another challenge may be the strong winds characteristic of the site, which may force delays and increase installation costs.

Exhibit 5 summarizes MDS's analysis. It envisions that a developer will purchase the Canyon Wind easement from WRI and immediately redevelop the property for sale to equity investors. The projected redevelopment includes replacing existing turbines with 20 new General Electric 1.5 megawatt models sporting 77-meter rotor diameters on 65-meter towers. It also anticipates negotiating a new 15-year, fixed-price PPA with Southern California Edison at 0.0588 $/kWh, and a minimum first-year interest coverage ratio of 1.75 times. Other assumptions are that the site's NCF will equal 43.74 percent, the salvage value of existing turbines will about equal the cost of removal, and interest rates on project debt will range between 6.5 and 7.0 percent.

The analysis indicates that the total value of the Canyon Wind Project at a PPA of $0.0588/kWh is $65.9 million. This number is driven by two key requirements: that equity investors see a prospective 15 percent IRR and that first-year interest coverage equals 1.75 times. Given these requirements, the spreadsheet in Exhibit 5 solves iteratively for total project value by calculating the present value of the equity investment and adding available debt financing. The project employs senior debt and PTC debt in a 2 to 1 ratio. Because creditors perceive PTC cash flows to be less risky than operating cash flows, the interest rate on a loan secured by PTC cash flows is lower than the rate available on senior debt.

With total development costs estimated to be $52.8 million, the implied developer profit is $13.0 million, well above the present value from continued operation of existing turbines. For comparison, MDS had assigned a value of $7.7 million to redevelopment of the same site in late 2003. Most of the increase was attributable to a 24 percent increase in the PPA, due in turn to rising natural gas prices.

MDS also prepared the matrix in Exhibit 6 showing the sensitivity of developer profit to 5 percent changes in the PPA price and the NCF. Exhibit 7 presents representative interest rates in July 2005.

The Decision

Two remaining issues puzzled Mr. Bittner as he reviewed MDS's report. Would a buyer pay the full developer profit calculated in Exhibit 5 to purchase the Canyon Wind easement, or in view of the risks surrounding redevelopment, would he only pay some fraction of this amount? And if so, what fraction should WRI expect? Redevelopment of the site certainly involved risk, but Mr. Bittner knew that due to the benefits of diversification only systematic, or nondiversifiable, risk should affect price. And to his eyes most of the risks associated with redeveloping the Canyon Wind easement appeared unsystematic.

In light of energy price volatility, Mr. Bittner also wondered if the ability to postpone redevelopment for at least three years might somehow contribute to the project's value in a way not captured in MDS's valuation. If so, MDS's estimated developer profit might understate true project value. Mr. Bittner thought in terms of a three-year horizon because the production tax credit was presently set to expire in three years, although Congress had repeatedly renewed the credit since 1992. Time was running short for a decision, and Mr. Bittner was anxious to get on with enjoying his summer.

Exhibit 1: Canyon Wind Farm Existing Turbines

Exhibit 2: Natural Gas Prices Monthly February (1985 – July 2005)

Cents/kWh
8.00
7.00
6.00
5.00
4.00
3.00
2.00
1.00
-
Nov-1984 Aug-1987 May-1990 Jan-1993 Oct-1995 Jul-1998 Apr-2001 Jan-2004

Exhibit 3: Volatility of Natural Gas Prices

Annualized Standard Deviation of Monthly Returns on US Natural Gas Wellhead Prices		
Date	Number of Observations	Volatility (%)
March 1985 – July 2005	245	35.1
Jan. 1995 – July 2005	127	41.7
Jan. 2000 – July 2005	67	44.1
Jan. 2003 – July 2005	31	43.0

Source: US Department of Energy, Energy Information Administration http://tonto.eia.doe.gov/dnav/ng/ng_pri_sum_dcu_nus_m.htm

Exhibit 4: Analysis of Continued Operation Using Existing Turbines

Price of electricity (per kWh)	$ 0.0588
Output (kWh/yr)	55,500,000
Investment in substation	$ 400,000
Net scrap value of a turbine in 2006	$ 1,200
Gross scrap value of turbine in 2016	$ 256
Inflation rate	2.5%
Discount rate	10%

Period	1	2	3	4	5	6	7	8	9	10
	2006	**2007**	**2008**	**2009**	**2010**	**2011**	**2012**	**2013**	**2014**	**2015**
Revenue	3,344,985	3,428,610	3,514,325	3,602,183	3,692,238	3,784,544	3,879,157	3,976,136	4,075,539	4,177,428
Costs										
Operations & Routine Maint.	1,107,500	1,135,188	1,163,567	1,192,656	1,222,473	1,253,035	1,284,360	1,316,469	1,349,381	1,383,116
Plant, substation & Edison fees	190,500	195,263	200,144	205,148	210,276	215,533	220,922	226,445	232,106	237,908
Land Rent	87,000	89,175	91,404	93,689	96,032	98,433	100,893	103,416	106,001	108,651
Insurance	240,000	246,000	252,150	258,454	264,915	271,538	278,326	285,285	292,417	299,727
Property tax	134,800	138,170	141,624	145,165	148,794	152,514	156,327	160,235	164,241	168,347
Management	106,100	108,753	111,471	114,258	117,115	120,042	123,043	126,120	129,273	132,504
Depreciation	200,000	200,000	200,000	200,000	200,000	200,000	200,000	200,000	200,000	200,000
Total	2,065,900	2,112,548	2,160,361	2,209,370	2,259,604	2,311,095	2,363,872	2,417,969	2,473,418	2,530,253
Pretax profit	1,279,085	1,316,062	1,353,964	1,392,813	1,432,633	1,473,449	1,515,285	1,558,167	1,602,121	1,647,174
Tax @ 40.7%	520,588	535,637	551,063	566,875	583,082	599,694	616,721	634,174	652,063	670,400
After tax profit	758,497	780,425	802,900	825,938	849,551	873,755	898,564	923,993	950,058	976,774
Depreciation	200,000	200,000	200,000	200,000	200,000	200,000	200,000	200,000	200,000	200,000
Cash flow from operations	958,497	980,425	1,002,900	1,025,938	1,049,551	1,073,755	1,098,564	1,123,993	1,150,058	1,176,774
Annual turbine overhaul	205,000	210,125	215,378	220,763	226,282	231,939	237,737	243,681	249,773	256,017
Free cash flow	753,497	770,300	787,522	805,175	823,270	841,817	860,827	880,313	900,285	920,758
Salvage value of turbines (after tax)										69,836
Land restoration cost (after tax)										-908,378
Total free cash flow	753,497	770,300	787,522	805,175	823,270	841,817	860,827	880,313	900,285	82,216
Present value (discounted at 10%)	**$4,715,520**									

Assumptions:

1. Output remains at 2005 level, provided $200,000 increasing at inflation rate is spent annually for major maintenance of turbines, in addition to routine maintenance.
2. Tax rates are 35% federal and 8.84% state (40.7% combined).
3. Restoration cost includes removal of substation and removal of old turbines, but not land restoration. Land restoration costs of about $1 million are deferred until the site is abandoned (perhaps in 2033). Turbine removal costs $2000 per unit. Net scrap value is the value after removal (i.e. Gross scrap value minus $2000).

Exhibit 5: Valuation of Canyon Wind Project Redevelopment by RHK Energy Consulting, Inc.

August 18, 2005

Site developed in 2006 with 20 Geneal Electric 1.5 megawatt SLE model turbines with 77 meter rotor diameters on 65 meter towers

Assumptions and Results	($ in thousands)
Capacity	
Turbine capacity (mw)	1.5
Number of turbines	20
Total capacity (mw)	30
Hours per year	8,760
Gross production/yr (mWh/yr)	262,800
Rated capacity factor	49.0%
Production before site adjustments	128,722
Site adjustments	13,768
Net adjusted annual production	114,954
Net capacity factor	43.74%
Development costs	
Equipment life (yr.s)	20
Salvage value in 20 yrs.	-
Cost per turbine & tower delivered	1,843
Total turbine & tower cost	36,860
Installation costs	8,060
Fees & expenses	7,921
Total development costs	$ 52,841
Power selling prices	
Purchase power agreement (yrs.)	15
PPA price ($/kWh)	0.0588
Sales in yrs. 16-20	at market
Increase in market price per year	2.5%

Financing			
Minimum IRR to equity		15%	
1st year interest coverage (times)		1.75	
1st year interest expense		2,669	
Debt sources			
	Rate	Term (yrs)	% Total debt
Senior debt	7.0%	15	67%
PTC debt	6.5%	10	33%
Weighted-average interest rate		6.835%	
Maximum debt		$ 39,043	
Senior debt		26,159	
PTC debt		12,884	
Compensating balance reqm't		2,366	
Tax rate (federal & state)		40.7%	
Depreciation		5 year MACRS	
Production tax credit rate		1.90%	
PTC COLA		2.05%	
Project Valuation			
Equity financing		26,829	
Senior debt financing		26,159	
PTC debt financing		12,884	
Total project value		65,871	
Developer profit		**$ 13,030**	

Contd...

Contd...

Cash flows to equity	0	1	2	3	4	5	6	7	8	9	10
Year	2006	2007	2008	2009	2010	2011	2012	2013	2014	2015	2016
Net production (MW)		114,949	114,949	114,949	114,949	114,949	114,949	114,949	114,949	114,949	114,949
PPA sales price ($/kWh)		0.0588	0.0588	0.0588	0.0588	0.0588	0.0588	0.0588	0.0588	0.0588	0.0588
Operating revenue ($ millions)		6,759	6,759	6,759	6,759	6,759	6,759	6,759	6,759	6,759	6,759
Other revenues		152	152	152	152	152	152	152	152	152	152
Total revenue		6,911	6,911	6,911	6,911	6,911	6,911	6,911	6,911	6,911	6,911
Total operating expenses		2,241	2,243	2,529	2,545	2,559	2,352	2,350	2,350	2,347	2,348
Operating income		4,670	4,668	4,382	4,366	4,352	4,559	4,561	4,561	4,564	4,563
Debt service		4,664	4,664	4,664	4,664	4,664	4,664	4,664	4,664	4,664	4,664
Pretax cash flow to equity		6	4	(282)	(298)	(312)	(105)	(103)	(103)	(100)	(101)
Tax calculation											
Operating income		4,670	4,668	4,382	4,366	4,352	4,559	4,561	4,561	4,564	4,563
Depreciation & Amort.		26,704	13,673	8,267	5,021	5,021	2,596	159	159	159	159
Interest expense		2,669	2,534	2,390	2,236	2,072	1,896	1,709	1,509	1,296	1,068
Taxable income		(24,703)	(11,538)	(6,275)	(2,891)	(2,741)	67	2,693	2,893	3,110	3,337
Tax		(10,054)	(4,696)	(2,554)	(1,177)	(1,116)	27	1,096	1,178	1,266	1,358
Production tax credit		2,184	2,229	2,275	2,321	2,369	2,417	2,467	2,517	2,569	2,622
FCF to equity		12,244	6,929	4,546	3,200	3,172	2,285	1,267	1,237	1,203	1,162
Equity investment for 15% IRR	**$ (26,829)**										
Depreciation calculations											
Asset basis		60,330	60,330	60,330	60,330	60,330	60,330	60,330			
MACRS depreciation rate		44.00%	22.40%	13.44%	8.06%	8.06%	4.04%				
Depreciation		26,545	13,514	8,108	4,863	4,863	2,437	0			
Amortizaztion calculations											
Asset basis		3,175	3,175	3,175	3,175	3,175	3,175	3,175	3,175	3,175	3,175
20 yr. SL amort.		159	159	159	159	159	159	159	159	159	159
Debt service calculations											
Sr. Debt interest		1,831	1,758	1,680	1,597	1,508	1,412	1,310	1,200	1,083	958
Sr. Debt principal pmt.		1,041	1,114	1,192	1,275	1,364	1,460	1,562	1,672	1,789	1,914
Sr. Debt Service		2,872	2,872	2,872	2,872	2,872	2,872	2,872	2,872	2,872	2,872
Ending Sr. Debt Principal		25,118	24,004	22,812	21,537	20,172	18,712	17,150	15,478	13,690	11,776
PTC Debt interest		837	775	709	639	564	484	399	309	212	109
PTC Debt principal pmt.		955	1,017	1,083	1,153	1,228	1,308	1,393	1,484	1,580	1,683
PTC Debt Service		1,792	1,792	1,792	1,792	1,792	1,792	1,792	1,792	1,792	1,792
Ending PTC Debt Principal		11,929	10,912	9,830	8,676	7,448	6,140	4,747	3,263	1,583	(0)
Production tax credit calculations											
Tax credit rate ($/KW)		0.0190	0.0194	0.0198	0.0202	0.0206	0.0210	0.0215	0.0219	0.0223	0.0228
Tax credit		2,184	2,229	2,275	2,321	2,369	2,417	2,467	2,517	2,569	2,622

Contd...

Contd...

Cash flows to equity	11	12	13	14	15	16	17	18	19	20
Year	*2017*	*2018*	*2019*	*2020*	*2021*	*2022*	*2023*	*2024*	*2025*	*2026*
Net production (MW)	114,949	114,949	114,949	114,949	114,949	114,949	114,949	114,949	114,949	114,949
PPA sales price ($/kWh)	0.0588	0.0588	0.0588	0.0588	0.0588	0.0852	0.0873	0.0895	0.0917	0.0940
Operating revenue ($ millions)	6,759	6,759	6,759	6,759	6,759	9,789	10,034	10,285	10,542	10,805
Other revenues	111	111	111	111	100	34	34	34	34	34
Total revenue	6,870	6,870	6,870	6,870	6,859	9,823	10,068	10,319	10,576	10,839
Total operating expenses	2,285	2,283	2,280	2,278	2,276	2,390	2,426	2,462	2,499	2,537
Operating income	4,585	4,587	4,590	4,592	4,583	7,433	7,642	7,857	8,077	8,302
Debt service	2,872	2,872	2,872	2,872	2,872	0				
Pretax cash flow to equity	1,713	1,715	1,718	1,720	1,711	7,433	7,642	7,857	8,077	8,302
Tax calculation										
Operating income	4,585	4,587	4,590	4,592	4,583	7,433	7,642	7,857	8,077	8,302
Depreciation & Amort.	159	159	159	159	159	159	159	159	159	159
Interest expense	824	681	528	363	188	0	0	0	0	0
Taxable income	3,602	3,747	3,904	4,070	4,236	7,274	7,483	7,698	7,918	8,143
Tax	1,466	1,525	1,589	1,656	1,724	2,961	3,046	3,133	3,223	3,314
Production tax credit	0	0	0	0	0	0	0	0	0	0
FCF to equity (pretax cf-tax)	247	190	129	64	(13)	4,472	4,596	4,724	4,854	4,988
Equity investment for 15% IRR										
Depreciation calculations										
Asset basis										
MACRS depreciation rate										
Depreciation										
Amortizaztion calculations										
Asset basis	3,175	3,175	3,175	3,175	3,175	3,175	3,175	3,175	3,175	3,175
20 yr. SL amort.	159	159	159	159	159	159	159	159	159	159
Debt service calculations										
Sr. Debt interest	824	681	528	363	188					
Sr. Debt principal pmt.	2,048	2,191	2,344	2,509	2,684					
Sr. Debt Service	2,872	2,872	2,872	2,872	2,872					
Ending Sr. Debt Principal	9,728	7,537	5,193	2,684	0					
PTC Debt interest										
PTC Debt principal pmt.										
PTC Debt Service										
Ending PTC Debt Principal										
Production tax credit calculations										
Tax credit rate ($/KW)										
Tax credit										

Contd...

Contd...

Discussion

Site adjustments. Efficiency adjustments to account for limited availability, electrical losses, wake and array losses, turbulence/high wind cut-out, blade contamination, icing, and grid outages.

Fees & expenses. Includes development expenses, capitalized interest during construction, capitalized term debt service reserve, lender's fee, lender's transaction costs, and borrower's counsel.

Purchase power agreement (PPA). A 15-year, fixed price power sales agreement we expect can presently be negotiated with Southern California Edison. The contact price depends primarily on the utility's highest power cost avoided by the contract, which for Southern California Edison is the cost of natural gas. We believe a contract can be negotiated today at 0.0588 $/kWh.

Sales in years 16-20. After 15 years, we assume power can be sold at a variable market price, which we estimate will increase with inflation at 2.5 percent a year.

Required IRR to equity. Based on our experience, we believe equity investors can be attracted to wind energy projects in today's markets that promise internal rates of returns of at least 15 percent.

Production tax credit rate (PTC). Congress offers production tax credits to encourage development of alternative energy. The current rate, recently extended for three years, is 1.90 percent of revenues increasing at 2.05 percent a year for 10 years.

Minimum equity/total capital. Based on experience and our market contacts we are confident this project can support a first-year interest coverage ratio as low as 1.75 times, with one-third of the debt secured by PTC cash flows. Because PTC cash flows depend only on revenue generation, lenders perceive them to be safer than operating cash flows and demand a lower interest rate. We estimate interest rate on the remaining debt to be 2 percent over 3-month LIBOR, or 7 percent.

Compensating balance requirement. Lenders demand that approximately one year's interest expense be held in reserve as a non-interest bearing deposit.

Depreciation. In addition to production tax credits Congress also allows rapid depreciation of wind energy projects. Even though the turbines have a 20-year life expectancy, ninety-five percent of total project value, less the compensating balance, qualifies for modified accelerated cost recovery (MACRS) depreciation over five years. The remaining 5 percent can be amortized on a straight-line basis over 20 years.

Developer profit. Equals the difference between Total project value and Total development costs.

Total operating expenses. Includes land lease payments, administrative expenses, property taxes, interconnection/wheeling expenses, insurance, and development royalties equal to 1.50 percent of gross revenue.

Equity investment for 15% IRR. Present value of free cash flows to equity through 2026 discounted at 15 percent.

Exhibit 6: Canyon Wind Project Sensitivity Analysis

Estimated Developer Profit ($ million)

5% Perturbations in PPA and NCF

PPA Price ($/kWh)

Net Capacity Factor	0.0559	*0.0588*	0.0617
41.55%	6.0	9.2	12.5
43.74%	9.6	13.0	16.5
45.93%	13.2	16.8	20.4

MDS Energy Consulting, Inc.

Exhibit 7: Representative Interest Rates in July 2005

Instrument	**Interest Rate (%)**
1-month Treasury Bill	3.10
3-month Treasury Bill	3.22
6-month Treasury Bill	3.28
1-year Treasury	3.64
3-year Treasury	3.91
5-year Treasury	3.98
10-year Treasury	4.18
20-year Treasury	4.48
5-year Treasury Inflation-Indexed	1.67
30-year Conventional Mortgage	5.70
BAA Corporate Bond Yield	6.25

(Robert C Higgins is Professor, Finance and Business Economics School of Business Administration, University of Washington. The author can be reached at rhiggins@u.washington.edu

Robert H Keeley Ph.D is Professor Emeritus at University of Colorado at Colorado Springs. The author can be reached at rkeeley@uccs.edu or rhkeeley@earthlink.net).

Annexure

ANNEXURE

Guidelines for Conducting Bird and Bat Studies at Commercial Wind Energy Projects

The 2002 New York State Energy Plan placed a priority on increased energy diversity and use of renewable energy sources, and the New York Renewable Portfolio Standard promotes the research and development of alternative energy projects, particularly commercial wind energy (*http://www.nyserda.org/rps/default.asp*). While wind energy has significant environmental benefits when compared to energy produced from fossil fuel, the Department of Environmental Conservation (DEC) must also consider the potential environmental impacts of wind energy production when evaluating proposed projects. Currently, the nature and severity of both site-specific and cumulative impacts that commercial wind energy projects may have on birds and bats in New York State is DEC's most pressing issue related to wind energy development.

These guidelines outline DEC's recommendations to commercial wind energy developers on how to characterize bird and bat resources at on-shore wind energy sites, and how to estimate and document impacts resulting from the construction and operation of projects. *Bird and bat resources* includes all species of birds (Class *Aves*) and bats (Order *Chiroptera*) that use or may use the site, as well as the habitats that support them. The guidelines also provide a general framework for the developer to propose site-specific studies needed to evaluate the potential and/or actual effects of a given wind energy project. It should be recognized that the effort required to fully understand the movement of birds and bats at any given site would be monumental and would take many years. Therefore, the studies recommended here are considered the minimum effort necessary to characterize bird and bat activity at a site within a reasonable time frame relative to project construction. This guidance provides for two tracks for pre-construction and post-construction studies: "standard" and "expanded." Many sites will warrant the "standard" studies. However, where site-specific conditions or findings suggest exceptional risk to birds and/or bats, expanded protocols and/or additional years of study will be recommended. Also, the protocols outlined are intended to provide comparability of data collection among sites and between years such that the information from each site may contribute to a statewide understanding of the ecological effects of wind energy generation.

1. Purpose

The purpose of this document is to outline the protocols for conducting bird and bat studies at wind energy projects to provide information necessary for DEC to:

a. Assess ongoing or expected environmental impact; and

b. Make a recommendation to the lead agency regarding the construction and operation of the project in order to avoid or minimize adverse environmental impact.

To perform such assessments and make a recommendation, DEC must consider information pertaining to the presence and activity of birds and bats at the site, or in the vicinity of the site. In this context, *the site* means not only the real property boundaries or outline of proposed turbine locations on the ground, but includes the air space over and surrounding the project. One of the most effective means of reducing direct and indirect impacts to birds and bats is to site turbines in a location that will cause the least disturbance to migrating, breeding, wintering, roosting, and feeding birds and bats. In addition to direct mortality from blade strikes, other negative effects from factors such as habitat loss or fragmentation, increased human activity and development, and increased predator presence can all result from the construction of a wind farm.

As wind energy development continues to expand throughout New York, more information is needed about the temporal and spatial use of habitats and the species composition of birds and bats using those habitats in order to relate wind energy production to its potential impacts. In particular, the recommendations for post-construction studies described in these guidelines are based on DEC's current knowledge of the best procedures for conducting thorough and meaningful post-construction mortality surveys at operating wind projects in New York. As post-construction mortality studies are conducted at more projects throughout the state over the next several years, these guidelines will be fine tuned to incorporate the most efficient, effective and accurate methodologies to fill post-construction data needs. Figure 1 illustrates the process described below for conducting pre- and post-construction studies.

2. Site and Project Description

A characterization of bird and bat resources includes documenting pertinent existing information, and collecting and analyzing additional data in the field on bird and bat use of the site. Few detailed studies have been conducted to date to gather site-specific data on where, when and how birds and bats use various habitats within the state. The study guidelines in this document are intended to provide a template for gathering such information and allowing for continued wind energy development that will not result in unacceptable impacts on the birds and bats of New York.

a. Compile Existing Information on Bird and Bat Resources

Prior to expending significant effort in planning a wind energy project, the developer should compile existing information on bird and bat resources at the site, including available information from other existing or proposed wind energy projects. The following sources should be consulted:

i. The DEC Central Office Division of Environmental Permits (*http://www.dec.ny.gov/about/642.html)* and Division of Fish, Wildlife and Marine Resources *(http://www.dec.ny.gov/about/634.html)* should be the initial point of contact for information regarding the permitting and assessment process for wind energy development;

ii. The New York Natural Heritage Program (NYNHP) should be contacted for information on state and federally listed species and sensitive ecological communities that may be located on or near the proposed project site *(http://www.dec.state.ny.us/website/dfwmr/heritage/);*

iii. Biologists in the DEC Regional office where the project is located should be contacted for available information on specific resources in the area of the proposed project site *(http://www.dec.ny.gov/about/255.html)*;

iv. Local ornithologists, Audubon Societies, birding clubs, hawk watches and nature centers can provide specific information about bird and bat resources, as well as reference to data from Breeding Bird Surveys, Breeding Bird Atlases and Christmas Bird Counts *(http://www.audubon.org/states/index.php?state=NY)*;

v. DEC's mammal specialist in the Bureau of Wildlife Endangered Species Unit can provide site specific information regarding the proximity of major bat hibernacula and summer roosting areas, as well as information on technical research being conducted within New York *(http://www.dec.state.ny.us/website/dfwmr/wildlife/endspec/)*; and,

vi. Bat Conservation International can provide general information about bats and bat biology (*http://www.batcon.org/home/default.asp*).

b. Identify Landscape Features and Resources of Concern

The presence of certain landscape features and/or ecological resources at a site can increase the likelihood that adverse impacts to bird and bat resources will result from a proposed wind energy project. The developer should identify whether any of these features or resources are present at the site of the proposed project. Such features or resources of concern include:

i. Habitat of a listed bird or bat species per 6 NYCRR Part 182 (e.g., species of special concern, threatened or endangered) (http://www.dec.ny.gov/animals/7494.html);

ii. Proximity of the project (approximately 5 miles) to the Atlantic coastline or the shoreline of one of the Great Lakes;

iii. The presence of or proximity to concentration areas of raptors, waterfowl or other vulernable species (approximately 2 miles) or a major bat hibernaculum (approximately 10 miles); and/or

iv. The presence of a specifically identified habitat or landscape feature (e.g., a ridgeline) that functions to funnel or concentrate birds or bats during migration or for feeding, breeding, wintering, roosting activities.

c. Provide Project Information to DEC

Once existing information is compiled, the developer should meet with DEC to discuss an overview of the proposal, the bird and bat resources of potential concern, and the application of these guidelines to the environmental assessment of the project.

To aid in project planning, the developer should prepare a complete description of the site, including:

i. Description of the geographical, topographical and other physical features of the site and within 10 miles of the site;

ii. Identification of state and federal wetlands, waterbodies, and drainage patterns;

iii. Location of permanent meteorological (met) towers, a summary of local weather patterns (e.g., annual precipitation, prevailing winds), and a summary of the wind resource; and,

iv. Maps with vegetation types, soils/bedrock, land use, and other information relevant to siting the project.

Prior to submitting the study work plan, additional information regarding the proposed project should be provided including:

v. Maps of the proposed turbine layout;

vi. Description of turbine type, size and rotor swept area;

vii. Figures showing existing and proposed roads, transmission line routes, and substation location.

Data regarding proposed site development should be provided in the form of shapefiles, coverages, geodatabases, and/or geometric networks for use in Geographical Information Systems (GIS) software including:

viii. Polygon coverages/shapefiles for the total project area as well as any concrete and building structures;

ix. Line coverages/shapefiles/geometric networks for the transmission and interconnect lines as well as proposed temporary construction and maintenance roads;

x. Polygons of the proposed temporary construction and maintenance roads for assessing the overall impact of the road footprints; and,

xi. Point coverages/shapefiles for any tower locations and/or any other structures that would be best represented as a point.

d. Prepare to Implement Standard or Expanded Study Protocols

Sites that contain, are within, or are in close proximity to the features or resources of concern listed in 2(b) above have the potential to cause unacceptable adverse impact on bird and bat resources. Therefore, projects proposed for such sites will require expanded rather than standard pre- and post-construction studies to identify and quantify potential or actual impacts. In particular, a proposal to site a wind energy project in proximity (10 miles) to an Indiana bat hibernaculum, along a coastline, or on a prominent ridgeline will result in a recommendation to conduct expanded pre-construction studies. In preparation for conducting either standard or expanded studies:

i. Contact the DEC Bureau of Fish and Wildlife Services Special Licenses Unit regarding necessary licenses/permits for collection and possession of birds and bats *(http://www.dec.ny.gov/permits/28633.html)* or special license to handle endangered species *(http://www.dec.ny.gov/permits/25012.html);*

ii. Contact the US Fish and Wildlife Service (USFWS) regarding Migratory Bird Treaty Act collection permits *(http://www.fws.gov/forms/3-200-7.pdf);*

iii. Engage an accredited wildlife biologist or ecologist knowledgeable about New York state fauna, natural history and habitat requirements; and experienced in wildlife study and habitat assessment protocols.

3. Study Objectives and Rationale

The overall goal of the recommended studies is to determine the potential for a specific wind energy project to have an adverse impact on bird and bat resources by characterizing the use of the site and surrounding area by bird and bat species under a variety of environmental conditions throughout the year, and by estimating the mortality rate of birds and bats due to collisions with turbines. The effects of construction and operation on habitat changes and changes in wildlife use of the project site will also be studied to determine any displacement or loss of species related to project operation. Data collected prior to construction can be compared to information collected in a similar matter after construction to determine what impact, if any, the project has on migrating and resident breeding and wintering birds and bats. Ultimately, information gained from pre- and post-construction studies will be used to identify mitigative measures that may be needed as part of an adaptive management plan to minimize direct and indirect impacts from project operation.

a. Pre-construction Studies

The objective of the pre-construction studies is to determine to what extent the proposed project area is used by migrating, breeding, and wintering birds and bats, and how the physical and biological features of the proposed site may influence such use.

b. Post-construction Studies

The objectives of the post-construction studies are:

i. To estimate direct impacts of the operating project in terms of mortality rates of birds and bats caused by collisions with turbines; and,

ii. To document any indirect impacts of construction and operation in the form of habituation/avoidance behavior of birds and bats in the area.

c. Bird Studies

Migrating birds, particularly neo-tropical migrants, are sensitive to changes occurring across the landscape that alter the amount and quality of habitat available to them during migration. Many aspects of the biology, population structure, and ecology of these birds are poorly understood. In a general sense, the following is known: most songbirds and many shorebirds and waterfowl migrate at night, while raptors move during the day; the exact spatial and temporal distribution of this migration is affected by weather patterns, food availability, and geographic features; concentrations of species and individual birds vary with the habitat, season, and year; birds are much more physiologically vulnerable during migration than at other times of the year; and the effects of human-caused habitat and landscape alterations are persistent over time.

Study methods for bird surveys include mist netting, reconnaissance surveys, habitat surveys for sensitive/listed species, and radar. The radar surveys provide information on target density, flight altitude, and flight direction. Acoustical monitoring of migratory birds can also be used to identify some species that vocalize in flight, and provide an estimate of flight height for these species. A combination of some or all of these methods will be recommended based on the specifics of the site, as each provides a different type and scope of information about the bird species utilizing the area.

d. Bat Studies

At this time, the greatest concern is for the Indiana bat and other bat species that typically migrate (red, hoary, silver-haired bats), as they are likely to be exposed to multiple wind facilities across much of their migration routes, and thus face the largest potential mortality. There is some evidence to suggest that pipistrelles are migratory as well. The timing of the fall migration extends into the end of October and the spring migration is underway by mid April. At this time, it is not known if these bats migrate across a broad front, if they use migratory corridors, or if their migration is affected by geographic features. However, it is assumed that the shores of the Great Lakes are more likely to have concentrations of migrating bats at lower altitudes than other regions of the state. Current methods for determining passage rates of bats and estimating the likelihood of collisions with turbines include mist netting, radar, thermal and/or light amplification imaging and acoustical monitoring. None of these methods provides definitive information and all have some drawbacks, although acoustical monitoring appears to hold the greatest promise because of the relatively low cost, low commitment of staff time, the ability to distinguish between birds and bats, and the ability to identify most individual calls to species.

4. Standard Pre-construction Studies

After compiling the site and project description and before commencing field studies, the developer should consult with DEC regarding the scope and specifics of pre-construction field studies at the project site. A minimum of one year of pre-construction studies is recommended at all proposed wind energy projects. Additional years of study will be recommended if warranted by the results of initial studies.

a. Weather Conditions

For all studies described in these guidelines (standard, expanded, pre- and post-construction) standard daily weather observations should be recorded any time field studies are being conducted. Weather information such as temperature, cloud cover, precipitation, wind speed and direction, and the timing of any cold or warm fronts passing through the area should be recorded on an hourly basis. Any additional weather information required for specific studies is identified in the individual study descriptions that follow.

b. Breeding Bird Surveys

Breeding bird surveys should be conducted in accordance with USGS methods during morning hours in June *(http://www.pwrc.usgs.gov/bbs/participate/instructions.html)*. The USGS method should be modified to include five-minute rather than three-minute counts to provide more thorough coverage. The number of survey points in each habitat type found within the project area should be representative of the number and location of turbines proposed in each habitat type. Each point should be surveyed by a single observer at least twice during the breeding season, with a minimum of seven days between surveys. All birds identified by sight or sound at each survey point should be recorded. Weather conditions should be conducive to hearing birdsong and seeing birds move about in vegetation and in flight. Excessively windy, rainy, or cold days should not be surveyed, as birds are not as active under these conditions. Observation points should be as close to proposed turbine locations as possible, and marked with GPS coordinates for future reference.

c. Habitat Surveys

Surveys should be undertaken at all project sites to identify existing habitat for New York State or federally listed rare, threatened or endangered species or State species of special concern If such habitat exists on or adjacent to the project area, additional surveys should be undertaken to determine if any such species are actually present on or near the project site. Developers should consult with DEC to determine the scope and timing of habitat surveys for a given species. Surveys should be seasonally appropriate for each of the listed target species, and their potential use of the area (e.g., summer for upland sandpiper, fall and spring for migrating golden eagles, and winter for short-eared owls).

d. Raptor Migration Surveys

Raptor migration surveys should be conducted from one or more prominent locations within the project area during spring and fall migration periods (April 1 to end of May; August 15 to November 1). Observations should take place between the hours of 9:00am and 4:00pm on at least 10 days during the predicted peak migration times for targeted species (see *http://hawkcount.org/index.php*), under the most favorable weather conditions (winds from the south during spring, winds from the north or northwest during fall). At a minimum, observations should include the peak two-week migration period for red-tailed, sharp-shinned and broad-winged hawks, and turkey vultures. It may be important to evaluate the migratory passage of other raptor species at site-specific locations as determined during consultation with DEC. Information on the species, number of individuals, sex and age class (if possible), behavior, flight height and direction, time of sighting, and location of each bird relative to the project area should be recorded. Concurrent with the information described above, observations of the movements of any other large flocks or individual birds (waterfowl, waders, corvids, icterids, swallows, etc.) should be recorded on a similar, separate data sheet.

e. Songbird Migration Surveys

Songbird migration surveys should be conducted for a minimum of once per week during the months of May and September. These surveys should be done from first light until no later than 11:00am on different days from the raptor surveys. The focus should be on songbirds, though other species, including soaring raptors and other fly-overs, should also be counted. Depending on the size and habitat distribution of the project area, one or more transects should be walked, with stops every 50 meters to record all species seen and heard during a 5 minute session. These surveys are intended to provide an estimate of the type and number of each species moving through the area in the spring and fall.

Conducting this survey separately from the raptor migration survey will allow for more time and attention to be given to detecting songbird species that move through the project area but do not nest or winter there, and would therefore likely be missed during other types of surveys. The location, length, and number of transects may vary for each project, and should be determined in consultation with DEC staff.

f. Bat Acoustical Monitoring

Movements of bats feeding at or passing through the site should be characterized using acoustical detectors. Detectors should be situated to sample no less than the rotor swept area, and to capture calls from as high an altitude as possible or at least 150 feet above ground surface. Wherever possible, detectors should be attached to existing meteorological (met) towers. As it is unlikely that there will be an adequate number of met towers to provide meaningful results, additional detector infrastructure will need to be installed. Two detectors sampling in a horizontal plane should be installed at each sampling location, one at roughly 22 meters above the ground and another at the top of the tower. Horizontal detectors should be oriented in the likely direction of arriving migrants (south in the spring, north in the fall). A vertical sampling detector should also be installed at the top of the tower. Recording should continue from 0.5 hrs prior to sunset until 0.5 hrs after sunrise daily between April 15 and October 15. Summer surveys should also include active acoustical sampling to determine which species are present on the site i.e., a field investigator with a detector walking across the study area in a variety of habitats that are likely to contain bats, and recording what is present. Active sampling should be conducted on at least nine warm (>55 deg F), dry, and calm evenings between June 1 and July 10, starting at dusk and ending no earlier than midnight. Analysis of calls should include the criteria used for species identification and should be verified by an independent authority.

5. Expanded Pre-construction Studies

If a developer proposes to construct a wind energy project in or near one of the features or resources of concern identified in section 2(b), two to three years of pre-construction study will be recommended incorporating one or more of the following expanded pre-construction studies to provide in-depth information on the bird and bat resources of the site.

a. Radar Studies

DEC will request the use of remote sensing marine radar to determine the use of the project area by nocturnally migrating birds and bats. The radar should sample in both horizontal and vertical modes to collect information on target density, flight height, direction, and speed. Radar units should be operated from sunset to sunrise April 15 to May 31 and August 15 to October 15. Data should be recorded in digital format. Nocturnal visual observations should be undertaken hourly during radar operation to estimate the proportion of birds and bats using the airspace immediately over or adjacent to the radar unit. Moon watching, spotlighting, and/or thermal imaging are the most commonly used methods. Consultation with DEC biologists is recommended to determine an appropriate location, duration, intensity, and time frame for these surveys.

b. Raptor Migration Surveys

For projects proposed to be sited on a ridgeline or in a known raptor migration route, expanded raptor migration surveys will be recommended. In addition, if observations during a standard study detected migrating raptor species listed by the state or federal government as threatened or endangered, expanded raptor surveys will be recommended. Surveys should be conducted from one or more prominent locations within the project area during spring and fall migration periods (April 1 to end of May; August 15 to November 1). If golden eagles are observed migrating through the project area, the fall observation period should extend through mid-December. Observations should take place between the hours of 9:00am and 4:00pm on as many days as possible during these predicted peak migration times for any targeted species, under the most favorable weather conditions (winds from the south during spring, winds from the north or northwest during fall). Information on the species, number of individuals, sex and age class (if possible), behavior, flight height and direction, time of sighting, and location of each bird relative to the project area should be recorded. Consultation with DEC biologists is recommended to determine an appropriate survey time frame for target species.

c. Waterfowl Surveys

Waterfowl surveys should be conducted if the project is in close proximity to a recognized major waterfowl concentration area. Surveys should include both driving and static observations in a variety of seasons and weather conditions. Driving surveys consist of slowly driving roads throughout the project site and surrounding area to observe and record the species, numbers, and behavior of birds in wetlands, rivers, fields and other habitats. For static surveys, an observer is stationed for a designated period of time at a given location making the same observations as driving surveys. Multiple static survey points should be located throughout the project area. Consultation with DEC biologists is recommended to determine an appropriate location, duration, intensity, and time frame for these surveys.

d. Breeding Bird Surveys

Targeted breeding bird surveys for species of concern should be conducted if the project is in close proximity to a wetland, grassland, forest or other habitat area that may harbor marshbirds, nightjars, owls, or other birds that would not easily be detected during a morning breeding bird survey, either because they are not active during the morning or are not typically vocal. The timing, duration, and method of detection for these surveys would be site-specific and dependant on the species involved. Consultation with DEC biologists is recommended to determine an appropriate location, duration, intensity, and time frame for these surveys.

e. Wintering Bird Surveys

Wintering bird surveys are recommended for projects that contains or are near a location known to harbor significant numbers of wintering birds, primarily focusing on but not limited to raptors. Particular attention should be paid to the presence of short-eared owls, snowy owls, northern harriers, bald eagles, and rough-legged hawks. Consultation with DEC biologists is recommended to determine appropriate location, duration, intensity, and time frame for these surveys.

f. Expanded Studies for Indiana Bats

If the project site is within 10 miles of an Indiana bat hibernaculum or known summer range, or if there is other information to suggest that Indiana bats are present, DEC will recommend an expanded version of the standard acoustical monitoring to include an inventory of the actual number of Indiana bats in the project area. The likelihood of bats from a maternity colony encountering a turbine diminishes as distance from the hibernaculum increases, with little chance of summering bats encountering a turbine more than 10 miles away. The Indiana bat inventory requires extensive mist netting with banding and radio tagging of captured bats as follows:

i. Bats should be captured in standard net sets, tagged and tracked to the roost;

ii. At each roost, bats leaving the roost should be captured, banded and radio tagged;

iii. This sequential tracking and netting process at each roost tree should be used to identify all of the maternity colonies in the project area;

iv. Concurrent exit counts of all bats leaving the identified roosts should be conducted to provide an estimate the number of individuals bats in the project area; and,

v. Exit counts of radio-tagged bats from each roost should be conducted on nights with sunset temperatures greater than 55 degrees F° over the duration of the transmitter's life (generally less than 20 days).

Wing punches and hair samples should be collected from all animals captured (regardless of species) and provided to DEC with appropriate chain of custody documentation. All bats captured should be fitted with a band provided by DEC. Any Indiana bat or silver-haired bat that is captured should be radio tagged and tracked to roost on a daily basis. DEC should be notified within 24 hrs of any captures of these two species. All information on capture (date, time, location etc.) and all radio tracking data for banded bats should be provided in a DEC specified database format.

g. Expanded Studies for Migratory Bats

The primary concern for migratory bats (red, hoary and silver-haired) is the mortality of migrants, especially migrants passing over waterbodies, because the potential for mortality is compounded as bats move seasonally through the many proposed turbine fields in eastern North America. In order to assess the potential mortality of migratory bats at a project proposed along a lakeshore, a network of acoustical sampling stations should be provided extending from the lakeshore to the portions of the proposed project area farthest from the lake such that variations in acoustical activity may be observed at the different sampling stations. The sampling should be conducted throughout the spring and fall migratory period from April 15 to October 15 and should include sampling at least to maximum turbine height. Consultation with DEC biologists is recommended to determine an appropriate location, duration, intensity, and time frame for these surveys.

In addition, to address the biases associated with acoustical sampling and validate the results of that sampling, a telemetry project should be conducted to locate the path of migrants through the project area. The recommended protocol is to:

i. Establish a line of automated receivers perpendicular to the assumed migration line, extending from the shoreline inland to the far periphery of the project;

ii. Based on receiver resolution, space receivers at intervals to ensure detection by at least two units for any migrating animal; and,

iii. Capture migrating bats at some distance (at least 25 miles) from the monitoring zone and attach transmitters.

Analysis of the telemetry data should assume that the movement patterns of a transmitting individual is representative of the species in terms of where they cross the monitoring zone.

6. Standard Post-construction Studies

The developer should conduct post-construction studies to evaluate actual impacts to birds and bats at the project site during turbine operation. Standard post-construction studies include mortality surveys, bird habituation and avoidance studies, and bat acoustical monitoring. DEC will evaluate the data from the first year of study to determine any changes to protocols that may become necessary after analysis and review of the initial data. The developer must ensure DEC staff and its agents full access to the site over the life of the project.

a. Ground Searches

Ground searches for bird and bat carcasses should be conducted under turbines at operating wind projects for a minimum of three years. For small projects (10 or fewer turbines), all the turbines should be searched. At larger projects (more than 10 turbines) at least 30% of the total number of turbines should be included in the ground searches.

i. **Turbine searches** – A standard turbine-searching pattern should be designed such that each turbine included in the study is searched either daily, weekly, or every three days from April 15 to November 15 during the first and second year after the entire project becomes operational. Whether the third year of study is done in sequence or postponed to a later year (e.g., the fifth) will be determined following analysis of data from the first and second years. Should the wind project expand to include more turbines, the number of turbines in the search pattern will be altered accordingly.

ii. **Area to be searched** –The area to be searched beneath each turbine should be no less than 1.5 times the rotor diameter. Transects should be approximately five (5) meters apart, allowing for a visual search area of 2.5 meters on either side of the centerline. These distances may vary slightly from one site to another, due to varying ground conditions.

iii. **Ground cover** – The type and amount of ground cover under each turbine should be recorded at regular intervals a minimum of once per week. Vegetation growth, crop harvesting and other changes in the substrate could greatly alter the efficiency of carcass recovery.

iv. **Search conditions** – Searches should begin as close to sunrise as possible. Overnight weather conditions greatly affect the number of animals that will fly and how they are distributed in the airspace, and thus their exposure to turbine blades. The standard weather data collection noted in section 4(a) need only be collected on a daily basis for ground searches.

v. **Photographs** – Digital photographs should be taken of each carcass found. At least one picture of each carcass should include a ruler or other standard item used for scale. These photos, along with all field data information described above, should be sent to DEC for species identification verification on a pre-arranged schedule. Photo documentation of each carcass should include:

(1) the position in which it was found;

(2) the dorsal and ventral sides;

(3) photos that indicate the gender and reproductive condition of bats (if possible);

(4) any identifying characteristics such as bill, foot, wing or tail shape, and plumage coloration for birds.

vi. **Data collection** – The following data should be recorded for each carcass found:

(1) Date, time, and turbine number;

(2) Location on plot marked with GPS coordinates,

(3) Distance and cardinal direction from turbine;

(4) Distance and bearing from transect from which it was first spotted;

(5) Condition of carcass (whole or partial, extent of injury and some measure of decomposition to estimate time of death);

(6) Position of carcass (face-up/down, sprawled, balled up, etc);

(7) Species, age and sex, if determinable;

(8) Substrate conditions when found (gravel, short/long grass, crops, brush, etc), and identification of collector.

b. Searcher Efficiency and Carcass Removal Trials

To more accurately estimate mortality rates, searcher efficiency tests and scavenger removal tests should be conducted throughout the study period using carcasses of various sizes and species. Factors such as ground topography, vegetation cover, current weather conditions, searcher experience and fatigue level, scavenging rates, and the use of dogs all affect the overall efficiency of carcass detection for a given project area. Methodologies for this type of study will evolve as new information is gathered, and specifics on data-gathering techniques will be updated and posted on the NYSDEC website. Searcher efficiency trials should be conducted to estimate search accuracy, and should take place unbeknownst to the searcher. The following process for conducting the trials is recommended:

i. **Carcass placement** – A project manager should place carcasses throughout the search areas under various turbines representing different types of ground cover early in the morning that a trial is to occur. The project manager should record the location of each carcass within the study area, and any not found by the searchers should be removed at the completion of the trial. Carcasses should be discreetly marked to identify them as test animals.

ii. **Carcass recovery** – Information collected on trial carcasses should be identical to all non-test carcasses as outlined in section 6(a)(vi). The number of test carcasses recovered and the accuracy of data recorded will be determined for each searcher, and an efficiency rate calculated for each trial conducted throughout the course of the study. The efficiency of any dogs used should be calculated separately from that of humans.

iii. **Carcass removal trials** – Carcass removal trials should take place periodically throughout the duration of the mortality study. Most mammalian and avian scavengers quickly recognize easy food sources, can readily incorporate wind farms in their daily routes, and are often active at pre-dawn hours. Insect scavengers are active mostly in warmer months, and in some cases can drastically deteriorate a carcass in a matter of hours. Carcass removal trials should continue throughout post-construction monitoring, as scavenging rates change in response to a steady source of food.

iv. **Number and condition of carcasses** – Carcasses should be as fresh as possible, since long-frozen carcasses are much more difficult to find and are far less attractive to scavengers. The number of carcasses used should not cause an excessive attraction to bring scavengers into the area. Carcasses should be placed in a variety of habitats and checked daily for the first week, and every two days thereafter until the carcass disappears (due to scavenging or decomposition). On each check, the location and condition of the carcass should be recorded to determine if any scavenging has occurred. Scavenging rates for each season and habitat type in the project area will be calculated.

c. Bird habituation and Avoidance Studies

The pre-construction Breeding Bird Survey should be repeated during the first and second years of operation. A third year of study should be conducted on the third, fourth or fifth year of project operation as determined by DEC. Post-construction survey points should be as close as possible to the location of pre-construction survey points. At pre-construction sample locations that become actual turbine sites, to the greatest extent possible, surveys should take place during a period when turbine noise does not interfere with the observer's ability to hear birds. Information from the post-construction BBS survey will examine whether the wind project is having any effect on resident/breeding bird use of the site, and whether habituation or avoidance is occurring.

d. Raptor Migration Surveys

The standard pre-construction raptor surveys should be repeated during the first year of operation. If dead or injured raptors are found under turbines during the first year of post-construction ground searches, then the expanded raptor surveys described below should be conducted in conjunction with subsequent years of mortality studies.

e. Bat Acoustical Sampling

Conduct standard bat acoustical monitoring following the same methods used for summer resident and migratory bat surveys during pre-construction, or per DEC recommendations. Detectors should be added to those wind turbines that will be monitored for kills on a daily basis. These detectors should be sampling the airspace above, below and perpendicular to the turbine hub.

f. Bat Specimen Collection

Entire bat carcasses should be collected and sent to DEC for species identification on a pre-arranged schedule. Should the developer decide not to provide the entire carcass, DEC will require a substantial (>1 cubic cm) sample of hair plucked from between the scapula, a full frame photograph of both the dorsal and ventral surface, and photographs that clearly indicate the sex and reproductive condition of both males and females. A complete wing and the head should also be collected and sent to DEC. All samples should be frozen immediately upon collection.

7. Expanded Post-construction Studies

For wind energy projects constructed in or near one of the identified features or resources of concern, expanded post-construction monitoring studies will be recommended to provide in-depth information on the impacts to bird and bat resources of the site.

a. Radar Surveys

If radar studies during pre-construction surveys showed high passage rates, low flight altitudes, or other unanticipated conditions were observed, then a radar survey will be recommended during the first year of post-construction mortality surveys. The use of radar during subsequent years of post-construction surveys will be contingent on the results of the first year of post-construction study. For any project where post-construction monitoring reveals a higher than expected level of mortality, the use of radar will be recommended for the following year of study regardless of whether radar studies were conducted during pre-construction studies.

b. Raptor Migration Surveys

Raptor surveys should be repeated during the first year of post-construction operation if expanded raptor surveys were conducted during pre-construction surveys. Raptor migration surveys should be done using the methods described under the expanded pre-construction survey section, or as recommended by DEC staff.

8. Bat Mortality and Genetic Isotope Analysis Project

As a means of assessing cumulative impacts to bat species from wind energy development in New York and along the Atlantic Coast, a regional bat mortality and genetic isotope analysis project is being implemented by a consortium of bat experts and resource agencies. The study will initially focus on the three New York bat species that appear most vulnerable to wind mortalities. These are the red *(Lasiurus borealis)* hoary (*Lasiurus cinereus*) and silver-haired (*Lasionycteris noctivagans*). The project will also determine if there are unique geographic population segments that should be of regional conservation concern. Bat mortality and genetic isotope studies should be conducted at each proposed and operating wind energy facility as follows:

a. Collect a minimum of 10 bat carcasses for genetic analysis over the duration of fall and spring migration and provide to DEC for collection of hair and tissue samples;

b. Provide funding for costs associated with the collection, field processing and analysis of specimens at approximately $250-$300 per animal; and

c. Repeat collection at a future date after build out of the wind resource in New York, to be coordinated with other regional collections.

9. Planning and Reporting

a. Work Plans

After discussions with DEC staff regarding application of these guidelines to a particular site, the developer should submit a draft work plan incorporating the recommended elements for study at the site. The work plan should include the site description and project layout provided for the initial consultation with staff. Pre-construction work plans should be discussed with and accepted by DEC before implementation of the proposed work. A comprehensive post-construction study plan should be developed and submitted to DEC for review and acceptance prior to project completion, and all work should be conducted in close consultation with DEC. Developers should work closely with DEC to provide a work plan detailing the search regime, bias corrections, reporting techniques, and other aspects of a project's post-construction mortality study.

b. Reports

After completion of the recommended studies, the developer should prepare a report presenting the results. A description of the proposed project should be provided including maps of the proposed turbine layout, existing and proposed roads, transmission line routes, substation location, topography, and state and federal wetlands. Maps with vegetation types, soils/bedrock, land use, significant archeological sites, and other relevant information should also be included. A composite map containing all project and study information (turbine locations, raptor observation points, BBS points, radar unit location (if applicable), wintering bird and waterfowl survey points/routes, and habitat types) should be provided. The preferred format for reporting is as follows:

i. **Breeding bird surveys:** The breeding bird survey report should identify the number of survey points, survey dates, the time and duration of surveys at each point, and the number of species identified at each point. A summary should include the number of species observed overall, species seen or heard most frequently throughout the study, point(s) with the greatest number of species, and habitat type(s) with the highest and lowest species diversity and abundance. The report should also include maps, tables and graphs reporting the location of each survey point and its relation to the proposed/existing turbine locations, and GPS coordinates of each point. Weather conditions during and immediately prior to survey days, a list of all species with the dates and points where they were observed, and the number and identification of observer(s) conducting each survey should also be provided.

ii. **Raptor migration survey:** The report should identify the number of observation point(s), the dates and times surveys were conducted, the number of species observed overall, species seen most frequently during the survey, the average and median flight height and direction of each species, and any notable behavior. The following information should also be included in the main report or as an appendix: the duration of each survey, the number and identification of observer(s) conducting each survey, a table containing an hourly breakdown of each survey day with information on weather conditions, the species composition, flight height and direction of each bird, a map with the observation point location(s) and overall mean flight paths over and adjacent to the project area.

iii. **Songbird migration surveys:** The report should identify the location and length of each transect, the number and location of survey points, survey dates, the time and duration of surveys at each point, and the number of species identified along each transect. Observations should be described as to number and behavior of birds seen e.g., solitary individual, moving in a small flock etc. The report should also include maps, tables and graphs reporting the location of each transect and its relation to the proposed/existing turbine locations, and GPS coordinates of each survey point along the transect(s). Weather conditions during and immediately prior to survey days, a list of all species with the dates and points where they were observed, and the number and identification of observer(s) conducting each survey should also be provided.

iv. **Habitat surveys:** The habitat survey report should describe the types of habitat found on site, and whether there are potentially any state and/or federally listed species that could be associated with each habitat type. The report should also include the results of listed species presence/absence surveys. A map of habitat locations (grassland, forest, shrub, wetland, etc.) within and near the project area should be provided, including the locations of habitat suitable for any listed species, as well as the locations of any actual observations made of listed species.

v. **Radar studies:** For each migration season, report the dates and total number of days that surveys took place; the radar unit location, elevation and surrounding vegetation/topography; and mean and median target flight height, direction, density (passage rate in targets/km/hr), and percentage of targets detected below the maximum height of the proposed turbines. The report should also include tables and graphs reporting the times and number of hours actually sampled each night, hourly weather information (particularly wind speed and direction, percent cloud cover, and the presence of fog and/or precipitation), mean and median hourly flight heights, direction, density and percentage of birds and bats below proposed turbine height. The evaluation of results should also report the amount of down time, failures, or suspected malfunctions. All performance data of equipment used should be reported to better assess the efficiency and accuracy of the units being used at each location.

vi. **Wintering bird surveys:** The report should include the number of survey points and/or routes surveyed, dates surveys were conducted, the duration and time of surveys at each point/location, the number of species observed overall, the species seen most frequently throughout the study, the point(s) with the greatest number of species, and the survey locations with the highest and lowest species diversity and abundance. Also included in the main report or as an appendix should description of the behavior (feeding, perching, soaring, flocking, etc.) of birds observed and the habitat they occupied, along with a map showing the locations of the sightings relative to proposed/existing turbine locations, weather conditions during and immediately prior to survey days, a list of all species with the dates and points where they were observed, and the number and identification of observer(s) conducting each survey.

vii. **Waterfowl surveys:** The static survey report should include the number of survey points, dates surveys were conducted, the duration and time of surveys at each point, the number of species observed overall, the species seen most frequently throughout the study, the point(s) with the greatest number of species, and the locations/habitat type (open water, river, marsh, agricultural field, etc.) with the highest and lowest species diversity and abundance. The driving survey report should include the same information provided for the static surveys, as well as the roads/areas that were driven/surveyed, and the type and number of species observed and their approximate location on the route. Behavior of individual birds and flocks seen should be noted (feeding, resting, flying, etc.), as well as any movements of birds within or across the project area. Also included in the main report or as an appendix should be maps, tables and/or graphs reporting the location of each survey point/route in relation to proposed/existing turbine locations, weather conditions during and immediately prior to survey days, a list of all species with the dates and points where they were observed, and the number and identification of observer(s) conducting each survey.

viii. **Bat acoustical surveys:** Acoustical data should be recorded and archived for quality control and to verify the identification of calls. Calls should be organized by species indicating the number of calls by date and by hour of the day. The report should identify the type of detectors in use, the performance data for each detector, its height above the ground, and its orientation. Include a description of the performance of each piece of equipment as it is configured for field data collection (sensitivity setting, housing, etc.) in order to determine the variability in detection based on species, and the distance from the target. Determine the reception range of all equipment used so as to identify variations between units and the variation between sensitivity settings used during the study. This should be described using a common metric (maximum detection range of each unit of a signal generated at a fixed DB at 20, 30 and 40 khz).

ix. **Mortality studies:** Details of the post construction mortality studies should be presented in a report consisting of the results of the ground searches with: the number of turbines searched and the frequency of searches (daily, weekly etc.), habitat type surrounding each turbine, table of how many birds and bats were found beneath each turbine, size class of carcass (small or large bat, small, medium or large bird), the dates found and condition of each carcass, daily weather conditions prior to and during each search, and the number of people and dogs, if applicable, conducting each survey. Results of the searcher efficiency tests and scavenger removal study should also be presented. The estimated searcher efficiency should be reported by carcass size and ground cover type for each searcher. Estimated scavenging rate should be reported for each carcass size, habitat type, and season. This should include the types of scavengers present on site (avian, mammalian, insect) and the frequency at which each occurs.

10. Sources of Information

Audubon of New York Christmas Bird Counts *http://www.audubon.org/chapter/ny/ny/christmas.html*

Audubon of New York Important Bird Areas *http://www.audubon.org/chapter/ny/ny/IBA_new.htm*

Audubon of New York Local Chapters *http://www.audubon.org/states/index.php?state=NY*

Bat Conservation International *http://www.batcon.org/home/default.asp*

DEC Breeding Bird Atlas *http://www.dec.state.ny.us/website/dfwmr/wildlife/bba/*

Hawk Migration Association of North America *http://hawkcount.org/index.php* Indiana Bat Recovery Plan *http://ecos.fws.gov/speciesProfile/SpeciesReport.do?spcode=A000*

New York Natural Heritage Program *http://www.dec.state.ny.us/website/dfwmr/heritage/*

NYSDEC Endangered Species Unit: contact Alan Hicks (bats) and Peter Nye (eagles) *http://www.dec.state.ny.us/website/dfwmr/wildlife/endspec/*

USFWS Migratory Bird Treaty Act Collection Permit *http://www.fws.gov/forms/3-200-7.pdf*

USGS Breeding Bird Survey *http://www.pwrc.usgs.gov/bbs/participate/instructions.html*

Figure 1: Flow Chart of Pre- and Post-Construction Studies

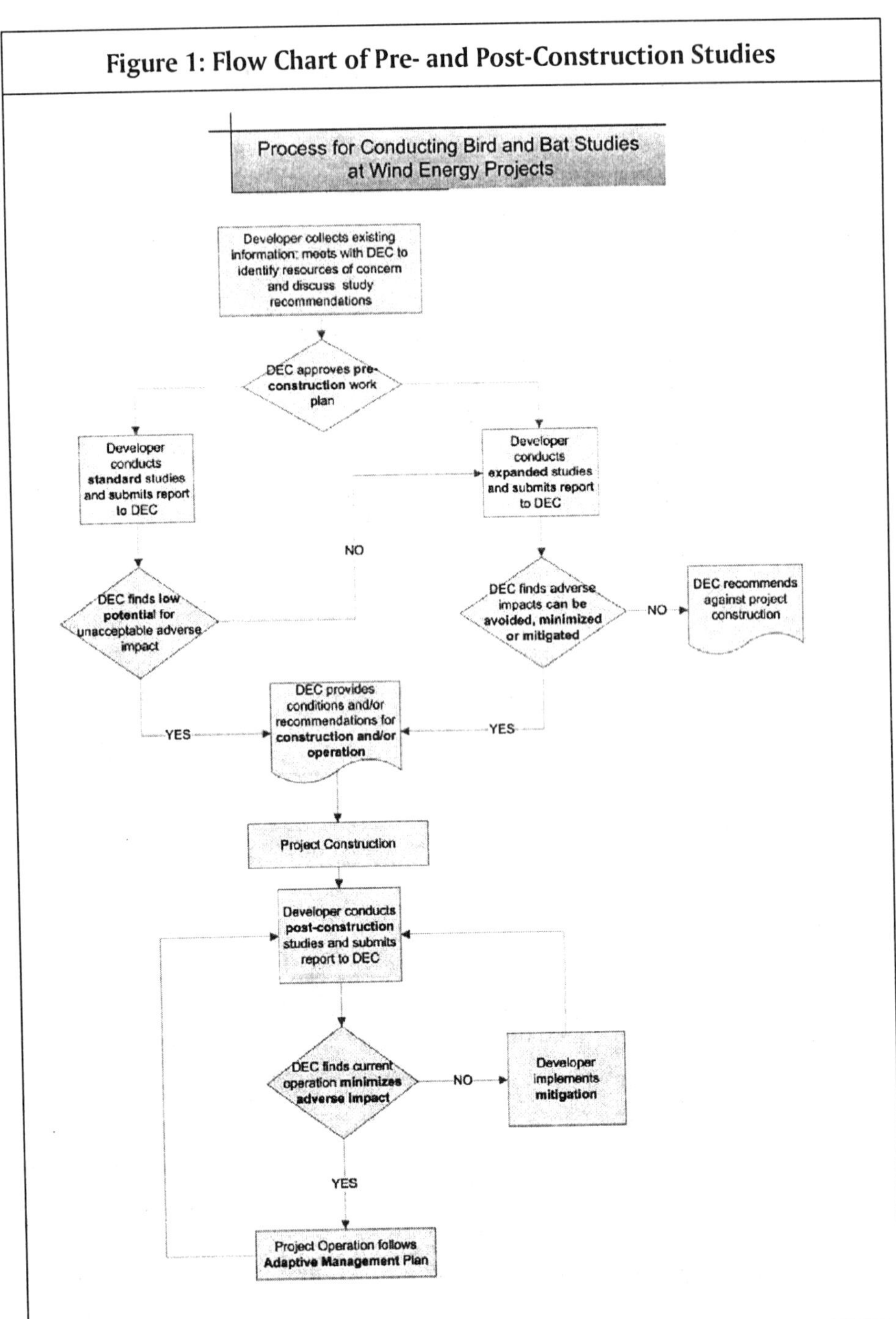

Index